The Microsoft® Excel Manual

Elementary Statistics

PICTURING THE WORLD

The Microsoft® Excel Manual

BEVERLY J. DRETZKE

Elementary Statistics

PICTURING THE WORLD

LARSON ▪ FARBER

PRENTICE HALL, Upper Saddle River, NJ 07458

Acquisitions Editor: Kathy Boothby Sestak
Supplement Editor: Joanne Wendelken
Special Projects Manager: Barbara A. Murray
Production Editor: Meaghan Forbes
Supplement Cover Manager: Paul Gourhan
Supplement Cover Designer: Liz Nemeth
Manufacturing Buyer: Alan Fischer

Upper Saddle River, NJ 07458

Printed in the United States of America

10 9 8 7 6 5 4 3 2 1

ISBN 0-13-015219-6

Prentice-Hall International (UK) Limited, London
Prentice-Hall of Australia Pty. Limited, Sydney
Prentice-Hall Canada, Inc., Toronto
Prentice-Hall Hispanoamericana, S.A., Mexico
Prentice-Hall of India Private Limited, New Delhi
Prentice-Hall (Singapore) Pte. Ltd.
Prentice-Hall of Japan, Inc., Tokyo
Editora Prentice-Hall do Brazil, Ltda., Rio de Janeiro

Introduction

The Microsoft® Excel Manual is one of a series of companion technology manuals that provide hands-on technology assistance to users of Larson/Farber *Elementary Statistics: Picturing the World.*

Detailed instructions for working selected examples, exercises, and Technology Labs from *Elementary Statistics: Picturing the World* are provided in this manual. To make the correlation with the text as seamless as possible, the table of contents includes page references for both the Larson/Farber text and this manual.

Contents:

Getting Started with Microsoft Excel

Overview

This manual is intended as a companion to Larson and Farber's *Elementary Statistics*. It presents instructions on how to use Microsoft Excel to carry out selected examples and exercises from *Elementary Statistics*.

The first section of the manual contains an introduction to Microsoft Excel and how to perform basic operations such as entering data, using formulas, saving worksheets, retrieving worksheets, and printing. All the screens pictured in this manual were obtained using the Office 97 version of Microsoft Excel on a PC. You may notice slight differences if you are using a different version or a different computer.

Getting Started with the User Interface

GS 1.1 The Mouse

The mouse is a pointer device that allows you to move around the Excel worksheet and to select specific locations and objects. There are four main mouse operations: Select, click, double-click, and right-click.

1. To **select** generally means to move the mouse pointer so that the white arrow is pointing at or is positioned directly over an object. You will often **select** commands in the standard toolbar located near the top of the screen. Some of the more familiar of these commands are open, save, and print.

2. To **click** means to press down on the left button of the mouse. You will frequently select cells of the worksheet and commands by "clicking" the left button.

3. To **double-click** means to press the left mouse button twice in rapid succession.

4. To **right-click** means to press down on the right button of the mouse. A right-click is often used to display special shortcut menus.

GS 1.2 The Excel Worksheet

The figure shown below presents the Office 97 version of a blank Excel worksheet. Important parts of the worksheet are labeled.

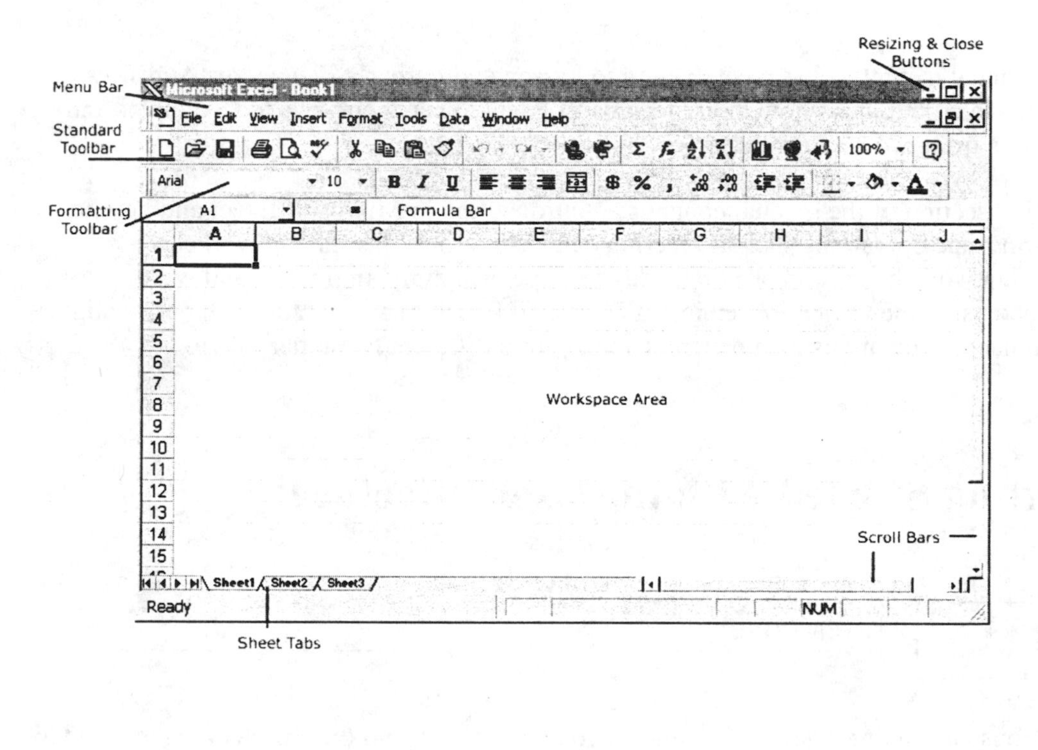

GS 1.3 Menu Conventions

Excel uses standard conventions for all menus. For example, the Menu bar contains the commands File, Edit, View, etc. Selecting one of these commands will "drop down" a menu. The Edit menu is displayed at the top of the next page.

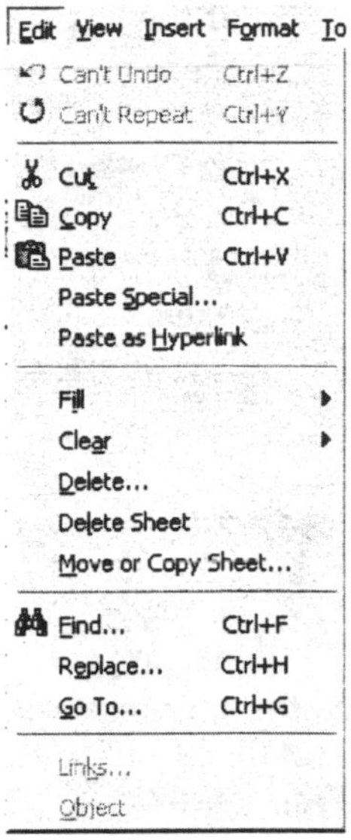

Icons to the left of the Cut, Copy, Paste, and Find commands indicate toolbar buttons that are equivalent to the menu choices.

Keyboard shortcuts are displayed to the right of the commands. For example, Ctrl+X is a keyboard shortcut for Cut.

The triangular markers to the right of Fill and Clear indicate that selection of these commands will result in a second menu of choices.

Selection of commands that are followed by an ellipsis (e.g., Paste Special… and Delete…) will result in the display of a dialog box that usually must be responded to in some way in order for the command to be executed.

The menus found in other locations of the Excel worksheet will operate in the same way.

GS 1.4 Dialog Boxes

Many of the statistical analysis procedures that are presented in this manual are associated with commands that are followed by dialog boxes. Dialog boxes usually require that you select from alternatives that are presented or that you enter your choices.

For example, if you select **Insert → Function**, a dialog box like the one shown below will appear. You are required to select both a Function category and a Function name. You make your selections by clicking on them.

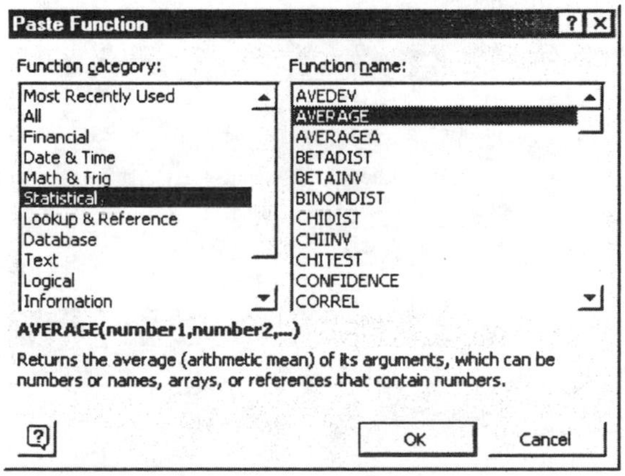

When you click the OK button at the bottom of the dialog box, another dialog box will often be displayed that asks you to provide information regarding location of the data in the Excel worksheet

Getting Started with Opening Files

GS 2.1 Opening a New Workbook

When you start Excel, the screen will open to **Sheet 1** of **Book 1**. Sheet names appear on tabs at the bottom of the screen. The name "Book 1" will appear in the top left corner.

If you are already working in Excel and have finished the analyses for one problem and would like to open a new book for another problem, follow these steps: First, at the top of the screen, click **File → New**. Next, click **OK** in the New dialog box. If you were previously working in Book1, the new worksheet will be given the default name Book 2.

The names of books opened during an Excel work session will be displayed at the bottom of the Window menu. To return to one of these books, click **Window** and then click the book name.

GS 2.2 Opening a File That Has Already Been Created

To open a file that you or someone else has already created, click on **File →Open**. A list of file locations will appear. Select the location by clicking on it. Many of the data files that are presented in your statistics textbook are available on the 3 ½ floppy disk that accompanies this manual. To open any of these files, you will select **3½ Floppy (A:)**.

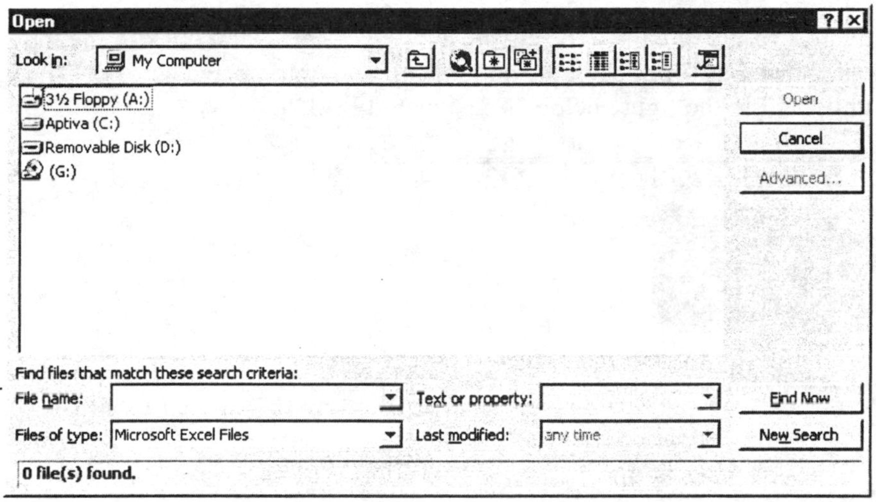

After you click on 3½ Floppy (A:), a list of folders and files available on the 3 ½ floppy disk will appear. You will need to select the folder or file you want by clicking on it. If you have selected a folder, another screen will appear with a list of files contained in the folder. Click on the name of the file that you would like to open.

Getting Started with Entering Information

GS 3.1 Cell Addresses

Columns of the worksheet are identified by letters of the alphabet and rows are identified by numbers. The cell address A1 refers to the cell located in column A row 1. The dark outline around a cell means that it is "active" and is ready to receive information. In the figure shown below, cell C1 is ready to receive information. You can also see C1 in the **Name Box** to the left of the **Formula Bar**. You can move to different cells of the worksheet by using the mouse pointer and clicking on a cell. You can also press [**Tab**] to move to the right or left, or you can use the arrow keys on the keyboard.

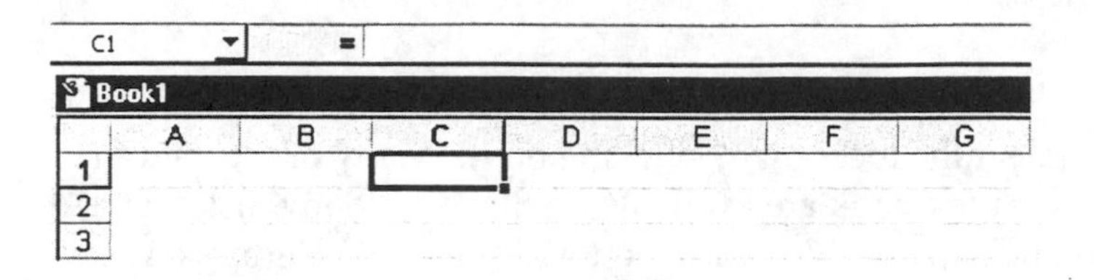

You can also activate a **range** of cells. To activate a range of cells, first click in the top cell and drag down and across (or click in the bottom cell and drag up and across). The range of cells highlighted in the figure below is designated B2:D6.

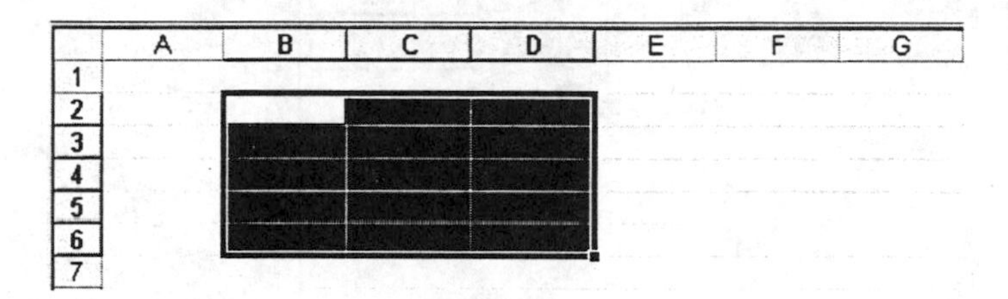

GS 3.2 Types of Information

Three types of information may be entered into an Excel worksheet.

1. **Text**. The term "text" refers to alphabetic characters or a combination of alphabetic characters and numbers, sometimes called "alphanumeric." The figure shown below provides an example of an entry comprised solely of alphabetic characters (cell A1) and an entry comprised of a combination of alphabetic characters and numbers (cell B1).

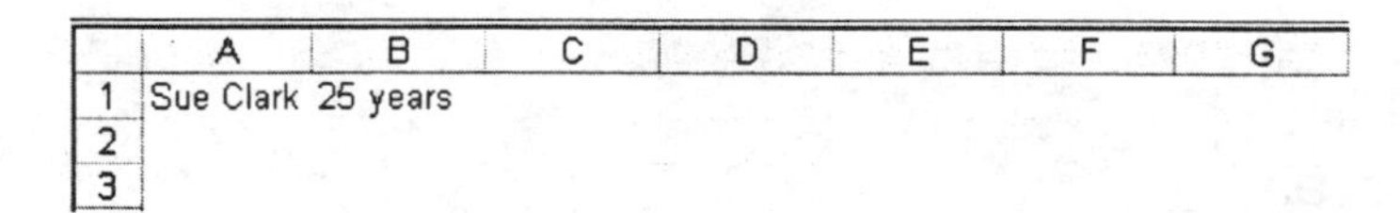

2. **Numeric**. Any cell entry comprised completely of numbers falls into the "numeric" category.

3. **Formulas**. Formulas are a convenient way to perform mathematical operations on numbers already entered into the worksheet. Specific instructions are provided in this manual for problems that require the use of formulas.

GS 3.3 Entering Information

To enter information into a cell of the worksheet, first activate the cell. Then key in the desired information and press [**Enter**]. Pressing the [Enter] key moves you down to the next cell in that column. The information shown below was entered as follows:

1. Click in cell A1. Key in **1**. Press [**Enter**].
2. Key in **2**. Press [**Enter**].
3. Key in **3**. Press [**Enter**].

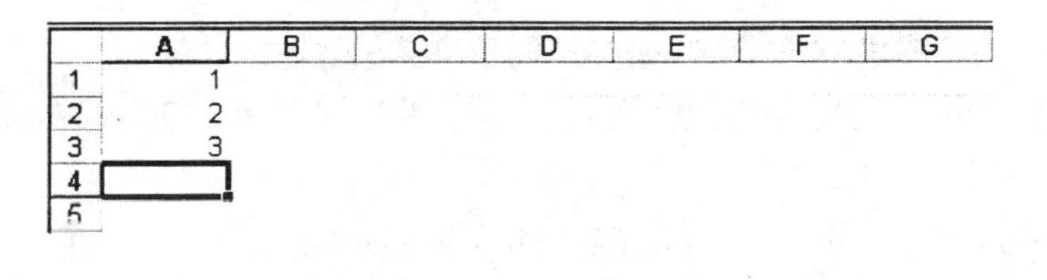

GS 3.4 Using Formulas

When you want to enter a formula, begin the cell entry with an equal sign (=). The arithmetic operators are displayed below.

Arithmetic operator	Meaning	Example
+	Addition	3+2
-	Subtraction	3-2
*	Multiplication	3*2
/	Division	3/2
^	Exponentiation	3^2

Numbers, cell addresses, and functions can be used in formulas. For example, to sum the contents of cells A1 and B1, you can use the formula =A1+B1. To divide by sum by 2, you can use the formula =(A1+B1)/2. Note that Excel carries out expressions in parentheses first and then uses the results to complete the calculations of the formula. Formulas will sometimes not produce the desired results because parentheses were necessary but were not used.

Getting Started with Changing Information

GS 4.1 Editing Information in the Cells

There are several ways that you can edit information that has already been entered into a cell.

1. If you have not completed the entry, you can simply backspace and start over. Clicking on the red X to the left of the Formula Bar will also delete an incomplete cell entry.

2. If you have already completed the entry and another cell is activated, you can click on the cell you want to edit and then press either [**Delete**] or [**Backspace**] to clear the contents of the cell.

3. If you want to edit part of the information in a cell instead of deleting all of it, follow the instructions provided in the example below.

    - Let's say that you wanted to enter 1234 in cell A1 but instead entered 124. Return to cell **A1** to make it the active cell by either clicking on it with the mouse or by using the arrow keys.

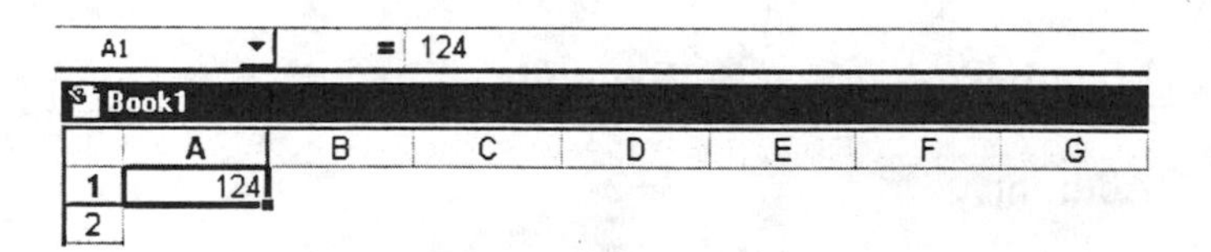

    - You will see A1 in the Name Box and 124 in the Formula Bar. Click between 2 and 4 in the Formula Bar so that the **I-beam** is positioned there. Enter the number 3 and press [**Enter**].

GS 4.2 Copying Information

To copy the information in one cell to another cell, follow these steps:

- First click on the source cell. Then, at the top of the screen, click **Edit →Copy**.
- Click on the target cell where you want the information to be placed. Then, at the top of the screen, click **Edit → Paste**.

To copy a range of cells to another location in the worksheet, follow these steps:

- First click and drag over the range of cells that you want to copy so that they are highlighted. Then, at the top of the screen, click **Edit →Copy**.
- Click in the topmost cell of the target location. Then, at the top of the screen, click **Edit →Paste**.

To copy the contents of one cell to a range of cells follow these steps:

- Let's say that you have entered a formula in cell C1 that adds the contents of cells A1 and B1 and you would like to copy this formula to cells C2 and C3 so that C2 will contain the sum of A2 and B2 and cell C3 will contain the sum of A3 and B3.

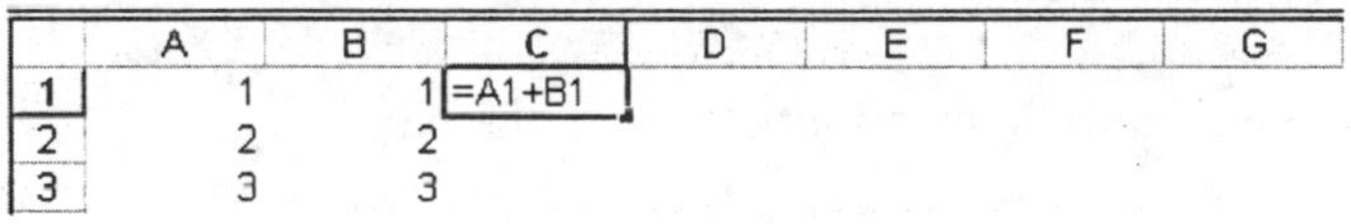

	A	B	C	D	E	F	G
1	1	1	=A1+B1				
2	2	2					
3	3	3					

- First click on cell C1 to make it the active cell. You will see =A1+B1 in the Formula Bar.

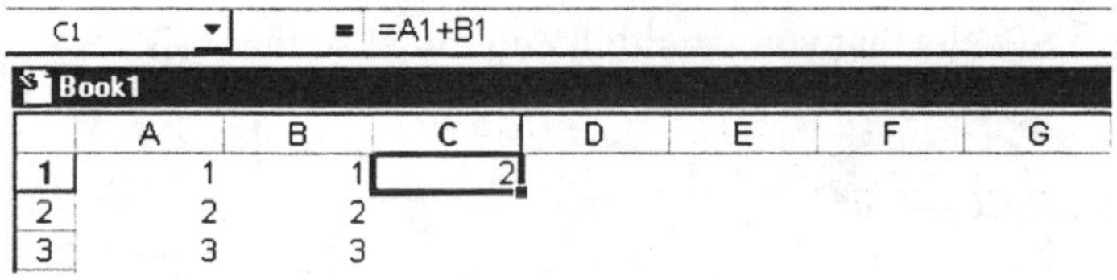

C1 = =A1+B1

Book1

	A	B	C	D	E	F	G
1	1	1	2				
2	2	2					
3	3	3					

- At the top of the screen, click **Edit → Copy**.

- Highlight cells C2 andC3 by clicking and dragging over them.

	A	B	C	D	E	F	G
1	1	1	2				
2	2	2					
3	3	3					

- At the top of the screen, click **Edit → Paste**. The sums should now be displayed in cells C2 and C3.

	A	B	C	D	E	F	G
1	1	1	2				
2	2	2	4				
3	3	3	6				

GS 4.3 Moving Information

If you would like to move the contents of one cell from one location to another in the worksheet, follow these steps:

- Click on the cell containing the information that you would like to move.
- At the top of the screen, click **Edit →Cut**.
- Click on the target cell where you want the information to be placed.
- At the top of the screen, click **Edit → Paste**.

If you would like to move the contents of a range of cells to a different location in the worksheet, follow these steps:

- Click and drag over the range of cells that you would like to move so that it is highlighted.
- At the top of the screen, click **Edit →Cut**.
- Click the topmost cell of the new location. (It is not necessary to click and drag over the entire range of the new location.)

- At the top of the screen, click **Edit** → **Paste**.

If you make a mistake, just click ***Edit*** *→* ***Undo****.*

GS 4.4 Changing the Column Width

There are a couple of different ways that you can use to change the column width. Only one way will be described here. Output from the Descriptive Statistics data analysis tool will be used as an example. As you can see in the output displayed below, many of the labels in column A can only be partially viewed because the column width is too narrow.

	A	B
1	*Test Score*	
2		
3	Mean	22.52632
4	Standard E	0.950126
5	Median	24
6	Mode	18
7	Standard [	4.141503
8	Sample Va	17.15205

Position the mouse pointer directly on the vertical line between A and B in the letter row at the top of the worksheet — A | B —so that it turns into a black plus sign. Click and drag to the right until you can read all the output labels. (You can also click and drag to the left to make columns narrower.) After adjusting the column width, you output should appear similar to the output shown below.

	A	B
1	*Test Score*	
2		
3	Mean	22.52632
4	Standard Error	0.950126
5	Median	24
6	Mode	18
7	Standard Deviation	4.141503
8	Sample Variance	17.15205

Getting Started with Sorting Information

GS 5.1 Sorting a Single Column of Information

Let's say that you have entered "Score" in cell A1 and four numbers directly below it and that you would like to sort the numbers in ascending order.

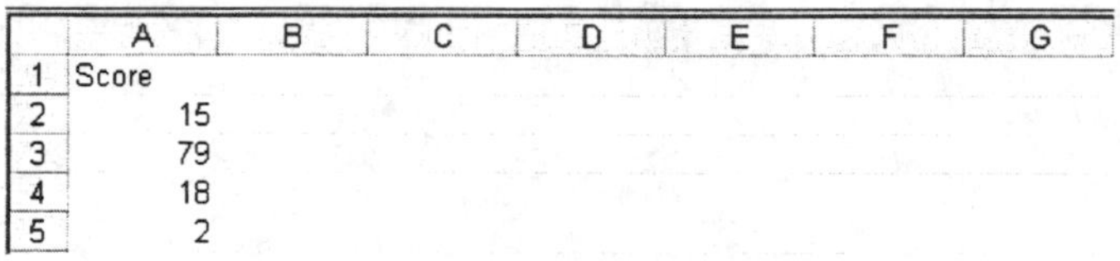

	A	B	C	D	E	F	G
1	Score						
2	15						
3	79						
4	18						
5	2						

- Click and drag from cell A1 to cell A5 so that the range of cells is highlighted.

You could also click directly on A in the letter row at the top of the worksheet. This will result in all cells of column A being highlighted.

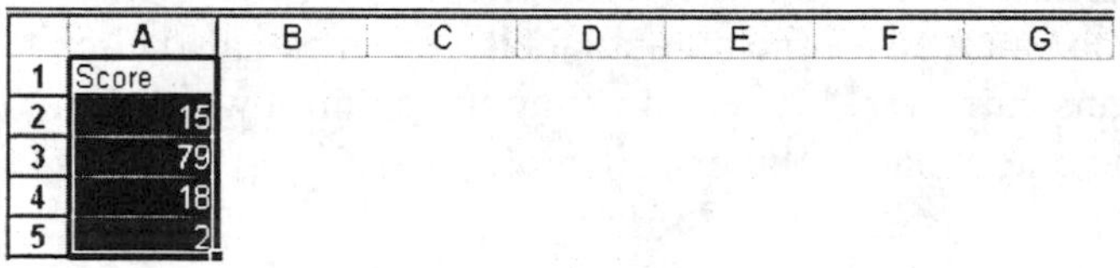

	A	B	C	D	E	F	G
1	Score						
2	15						
3	79						
4	18						
5	2						

- At the top of the screen, click **Data → Sort**.
- In the Sort dialog box that appears, you are given the choice of sorting the information in column A in either ascending or descending order. The ascending order has already been selected. Header row has also been selected. This means that

the "Score" header will stay in cell A1 and will not be included in the sort. Click **OK** at the bottom of the dialog box.

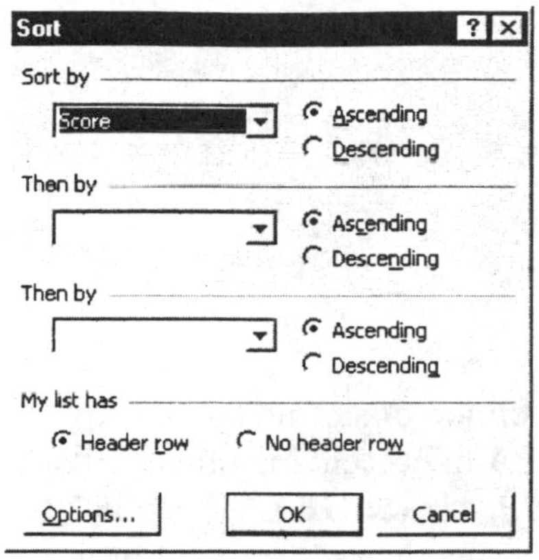

Sort
Sort by
Score — (•) Ascending ( ) Descending
Then by
(•) Ascending ( ) Descending
Then by
(•) Ascending ( ) Descending
My list has
(•) Header row ( ) No header row
Options... OK Cancel

The cells in column A should now be sorted in ascending order as shown below.

	A	B	C	D	E	F	G
1	Score						
2	2						
3	15						
4	18						
5	79						

GS 5.2 Sorting Multiple Columns of Information

Your Excel data files will frequently contain multiple columns of information. When you sort multiple columns at the same time, Excel provides a number of options.

Let's say that you have a data file that contains the information shown below and that you would like to sort the file by GPA in descending order.

	A	B	C	D	E	F	G
1	Score	Age	Major	GPA			
2	2	19	Music	3.1			
3	15	19	History	2.4			
4	18	22	English	2.7			
5	79	20	English	3.7			

- Click and drag from A1 down and across to D5 so that the entire range of cells is highlighted.

	A	B	C	D	E	F	G
1	Score	Age	Major	GPA			
2	2	19	Music	3.1			
3	15	19	History	2.4			
4	18	22	English	2.7			
5	79	20	English	3.7			

- At the top of the screen, click **Data** → **Sort**.

- In the Sort dialog box that appears, you are given the option of sorting the data by three different variables. You want to sort only by GPA in descending order. Click the down arrow to the right of the Sort by window until you see **GPA**. Then click the button to the left of **Descending** so that a black dot appears there. You want the variable labels to stay in row 1, so **Header row** should be selected. Click **OK**.

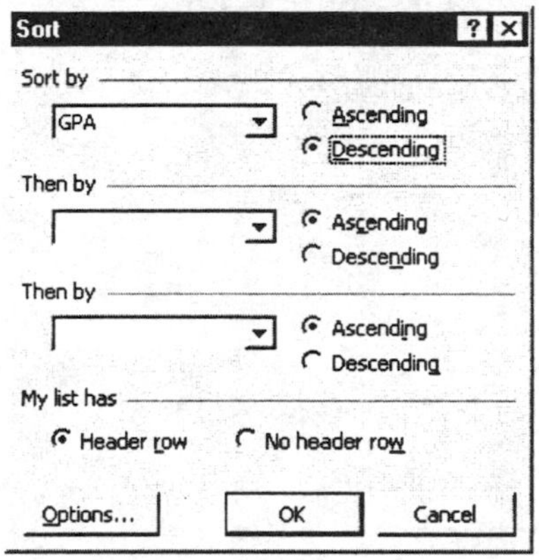

Sort ? X
Sort by
GPA
Ascending
Descending
Then by
Ascending
Descending
Then by
Ascending
Descending
My list has
Header row No header row
Options... OK Cancel

The sorted data file is shown below.

	A	B	C	D	E	F	G
1	Score	Age	Major	GPA			
2	79	20	English	3.7			
3	2	19	Music	3.1			
4	18	22	English	2.7			
5	15	19	History	2.4			

Getting Started with Saving Information

GS 6.1 Saving Files

To save a newly created file for the first time, click **File → Save** at the top of the screen. A Save As dialog box will appear. You will need to select the location for saving the file by clicking on it. In the dialog box shown below, the 3½ Floppy has been selected.

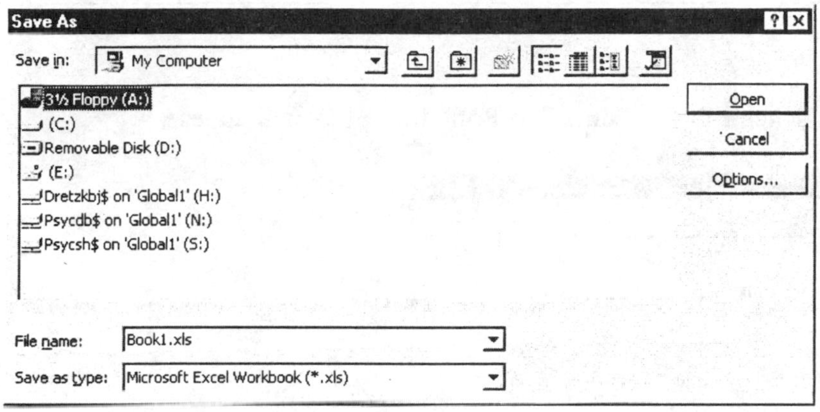

The default file name, displayed in the File name window, is **Book1.xls**. It is highly recommended you replace the default name with a name that is more descriptive. It is also highly recommended that you use the **xls** extension for all your Excel files.

Once you have saved a file, clicking **File → Save** will result in the file being saved in the same location under the same file name. No dialog box will appear. If you would like to save the file in a different location, you will need to click **File → Save As**.

GS 6.2 Naming Files

Windows 98 and Mac versions of Excel will allow file names to have around 200 characters. The extension can have up to three characters. You will find that long, descriptive names will be easier to work with than really short names. For example, if a file contains data that was collected in a survey of Milwaukee residents, you may want to name the file **Milwaukee survey.xls**.

Several symbols cannot be used in file names. These include: forward slash (/), backslash (\), greater-than sign (>), less-than sign (<), asterisk (*), question mark (?), quotation mark ("), pipe symbol (|), color (:), and semicolon (;).

Getting Started with Printing Information

GS 7.1 Printing Worksheets

To print a worksheet, click **File → Print**. The Print dialog box will appear.

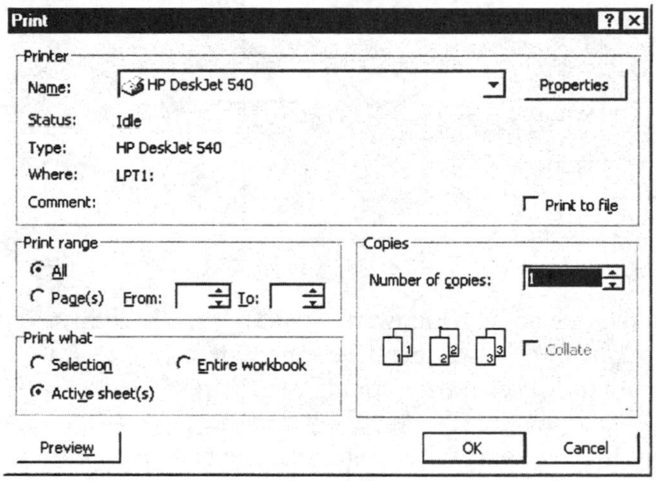

Under Print range, you will usually select **All**, and under Print what, you will usually select **Active sheet(s)**. The default number of copies is 1, but you can increase this if you need more copies. When the Print dialog box has been completed as you would like, click **OK**.

GS 7.2 Page Setup

Excel provides a number of page setup options for printing worksheets. To access these options, click **File → Page Setup**. Under **Page**, you may want to select the Landscape orientation for worksheets that have several columns of data. Under **Sheet**, you may want to select Gridlines.

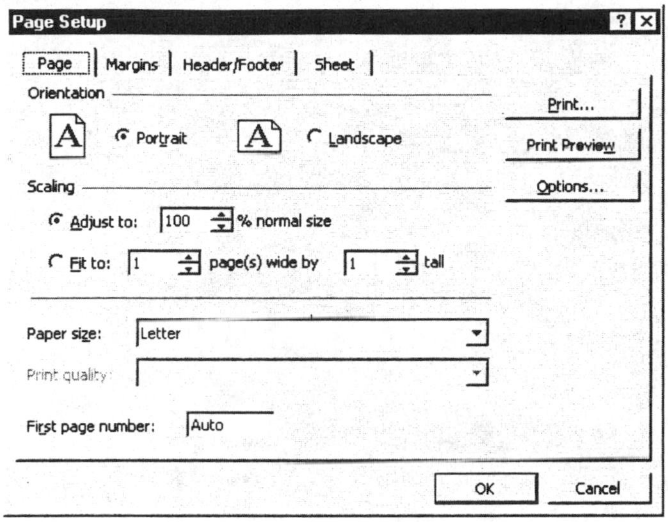

Getting Started with Add-ins

GS 8.1 Loading Excel's Analysis Toolpak

Thc Analysis Tookpak is an Excel Add-In that may not necessarily be loaded. If it does not appear at the bottom of the Tools menu, then click on **Add-Ins** in the Tools menu to get the dialog box shown below.

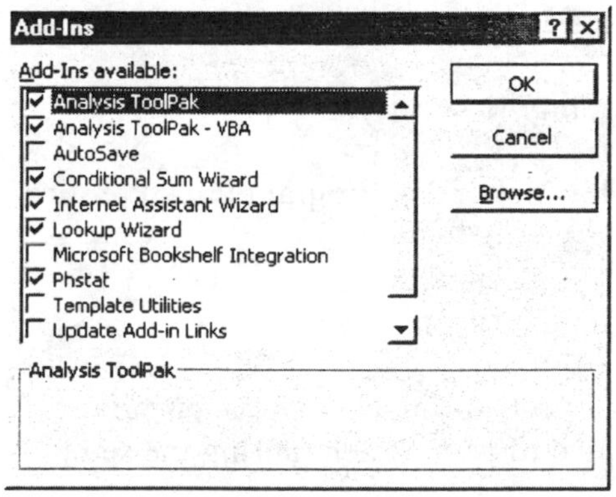

Click in the box to the left of **Analysis ToolPak** to place a checkmark there. Then click **OK**. The ToolPak will load and will be listed at the bottom of the Tools menu as shown below.

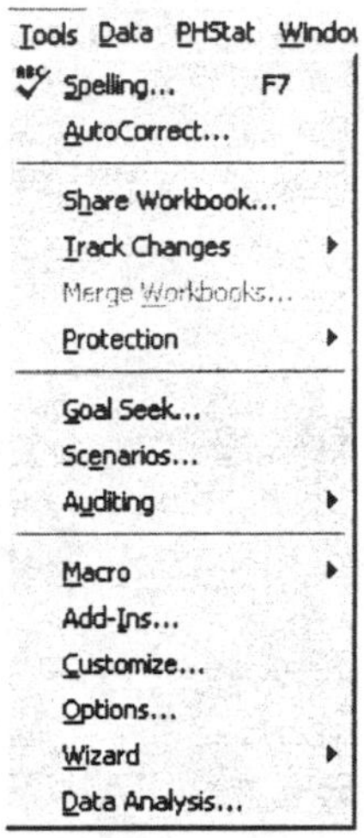

Tools Data PHStat Windo
Spelling... F7
AutoCorrect...
Share Workbook...
Track Changes
Merge Workbooks...
Protection
Goal Seek...
Scenarios...
Auditing
Macro
Add-Ins...
Customize...
Options...
Wizard
Data Analysis...

GS 8.2 Loading the PHStat Add-In

PHStat is a Prentice Hall statistical add-in that is included on the CD-ROM that accompanies your statistics textbook. The instructions that are given here also appear in the PHStat readme file.

To use the Prentice Hall PHStat Microsoft Excel add-in, you first need to run the setup program (Setup.exe) located in the PHStat directly on the disk. The setup program will install the PHStat program files to your system and add icons on your Desktop and Start Menu for PHStat. Depending on the age of your Windows system files, some Windows system files may be updated during the setup process as well.

During the Setup program you will have the opportunity to specify:

a. The directory into which to copy the PHStat files (default is \Program Files\Prentice Hall\PHStat).

b. The name of the Start Programs folder to contain the PHStat icon.

To begin using the Prentice Hall PHStat Microsoft Excel add-in, click the appropriate Start Menu or Desktop icon for PHStat that was added to your system during the setup process. Select the PHStat for Excel 95 if your system has Excel 95 installed or select PHStat for Excel 97 if your system has Excel 97 installed.

When a new, blank Excel worksheet appears, check the Tools menu to make sure that both the **Analysis ToolPak** and **Analysis ToolPak–VBA** have been checked.

Introduction to Statistics

CHAPTER 1

Technology Lab

▸ Example (pg. 22)

Generating a List of Random Numbers

You will be generating a list of 15 random numbers between 1 and 167 to use in selecting a random sample of 15 cars assembled at an auto plant.

If the PHStat add-in has not been loaded, you will need to load it before continuing. Follow the instructions in Section GS 8.2.

---

1. Open a new, blank Excel worksheet.

2. At the top of the screen, select **PHStat → Data Preparation → Random Sample Generator**.

3. Complete the Random Sample Generator dialog box as shown below. A sample of 15 cars will be randomly selected from a population of 167. The topmost cell in the output will contain the title "Car #." Click **OK**.

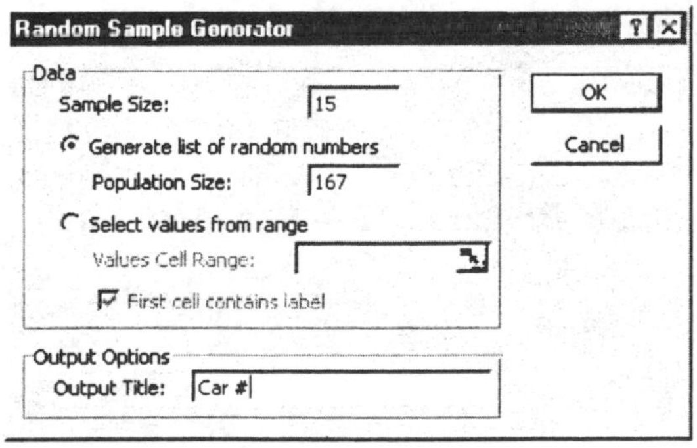

Random Sample Generator

Data
Sample Size: 15
Generate list of random numbers
Population Size: 167
Select values from range
Values Cell Range:
First cell contains label
Output Options
Output Title: Car #
OK
Cancel

The output is displayed in a new worksheet named "RandomNumbers." Because the numbers were generated randomly, it is not likely that your output will be exactly the same.

	A	B	C	D	E	F	G
1	Car #						
2	119						
3	20						
4	9						
5	35						
6	82						
7	45						
8	39						
9	148						
10	127						
11	58						
12	47						
13	156						
14	14						
15	34						
16	123						

◄

► Exercise 1 (pg. 23) Generating a List of Random Numbers for a CPA

You will be generating a list of 8 random numbers between 1 and 74 to use in selecting a random sample of accounts. You will also be ordering the generated list from lowest to highest.

If the PHStat add-in has not been loaded, you will need to load it before continuing. Follow the instructions in Section GS 8.2.

1. Open a new, blank Excel worksheet.

2. At the top of the screen, select **PHStat → Data Preparation → Random Sample Generator**.

3. Complete the Random Sample Generator dialog box as shown below. A sample of eight accounts will be randomly selected from a population of 74. "Account #" will appear in the top cell of the output. Click **OK**.

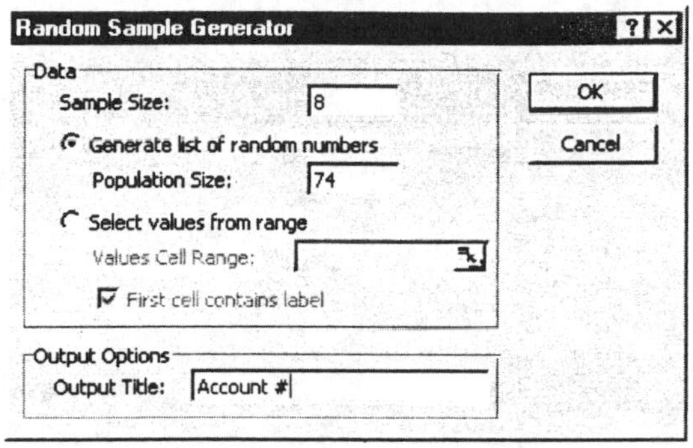

Random Sample Generator

Data
Sample Size: 8
Generate list of random numbers
Population Size: 74
Select values from range
Values Cell Range:
First cell contains label
OK
Cancel

Output Options
Output Title: Account #

4. Sort the numbers in ascending order.

For instructions on how to sort, refer to Sections GS 5.1 and GS 5.2.

The sorted set of eight random numbers is displayed below. Because the numbers were generated randomly, it is not likely that your output will be exactly the same.

	A	B	C	D	E	F	G
1	Account #						
2	13						
3	20						
4	32						
5	43						
6	57						
7	58						
8	59						
9	68						

◄

► Exercise 2 (pg. 23) Generating a List of Random Numbers for a Quality Control Department

If the PHStat add-in has not been loaded, you will need to load it before continuing. Follow the instructions in Section GS 8.2.

1. Open a new Excel worksheet.

2. At the top of the screen, select **PHStat → Data Preparation → Random Sample Generator**.

3. Complete the Random Sample Generator dialog box as shown below. A sample of 20 batteries will be randomly selected from a population of 200. "Battery #" will appear in the top cell of the generated output. Click **OK**.

Random Sample Generator

Data
Sample Size: 20
Generate list of random numbers
Population Size: 200
Select values from range
Values Cell Range:
First cell contains label

Output Options
Output Title: Battery #

OK
Cancel

4. Sort the numbers in ascending order.

For instructions on how to sort, see Sections GS 5.1 and GS 5.2.

The sorted set of 20 random numbers is displayed below. Because the numbers were generated randomly, it is unlike that your output will be exactly the same.

	A	B	C	D	E	F	G
1	Battery #						
2	1						
3	5						
4	11						
5	50						
6	58						
7	59						
8	62						
9	65						
10	76						
11	85						
12	91						
13	93						
14	111						
15	117						
16	148						
17	155						
18	165						
19	168						
20	182						
21	186						

◀

► Exercise 3 (pg. 23) Generating Three Random Samples from a Population of Ten Digits

You will be generating three random samples of five digits each from the population: 0, 1, 2, 3, 4, 5, 6, 7, 8, 9. You will also compute the average of each sample.

If the PHStat add-in has not been loaded, you will need to load it before continuing. Follow the instructions in Section GS 8.2.

1. Open a new Excel worksheet. Enter the numbers 0 through 9 in column A.

	A	B	C	D	E	F	G
1	0						
2	1						
3	2						
4	3						
5	4						
6	5						
7	6						
8	7						
9	8						
10	9						

2. Compute the average using the AVERAGE function. To do this, first click in the cell immediately below the last number in the population—cell A11. Then, at the top of the screen click **Insert → Function**.

3. Under Function category, select **Statistical**. Under Function name, select **AVERAGE**.

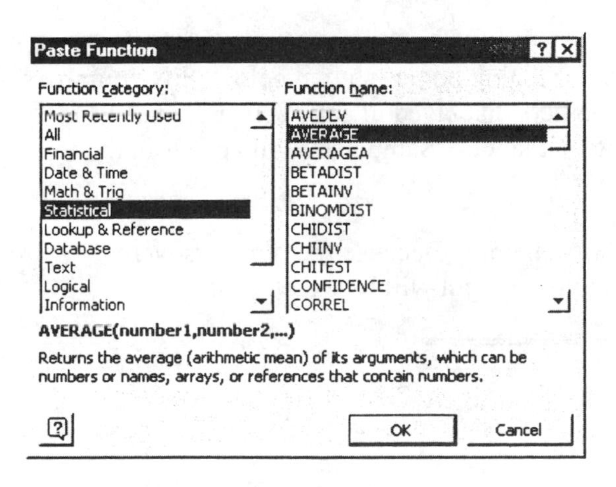

4. Complete the AVERAGE dialog box as shown below. Click **OK**.

A quick way to enter the range is to first click in the Number 1 field in the dialog box. Then click on cell A1 in the worksheet and drag down to cell A10.

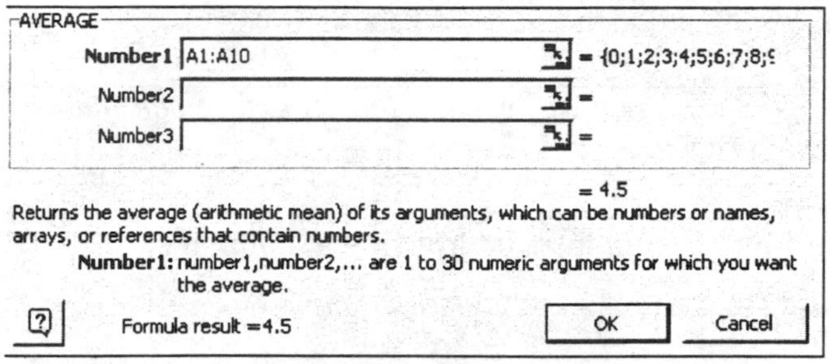

AVERAGE

Number1 A1:A10 = {0;1;2;3;4;5;6;7;8;9

Number2 =

Number3 =

= 4.5

Returns the average (arithmetic mean) of its arguments, which can be numbers or names, arrays, or references that contain numbers.

Number1: number1,number2,... are 1 to 30 numeric arguments for which you want the average.

Formula result =4.5 OK Cancel

The average, 4.5, is displayed in cell A11 of the worksheet.

	A	B	C	D	E	F	G
1	0						
2	1						
3	2						
4	3						
5	4						
6	5						
7	6						
8	7						
9	8						
10	9						
11	4.5						

5. At the top of the screen, select **PHStat → Data Preparation → Random Sample Generator**.

6. Complete the Random Sample Generator dialog box as shown below. Five numbers will be randomly selected from the numbers displayed in cells A1 through A10 in the worksheet. Cell A1 does not contain a label. "Sample 1" will appear in the top cell of the generated output. Click **OK**.

The random sample of five digits is shown below. Because the numbers were generated randomly, it is not likely that your output will be exactly the same.

	A	B	C	D	E	F	G
1	Sample 1						
2	6						
3	5						
4	0						
5	9						
6	2						

7. Use the AVERAGE function to find the average. To do this, first click in the cell immediately below the last number in the sample—cell A7. Then, at the top of the screen, click **Insert →Function**.

8. Under Function category, select **Statistical**. Under Function name, select **AVERAGE**. Complete the AVERAGE dialog box as shown below. Click **OK**.

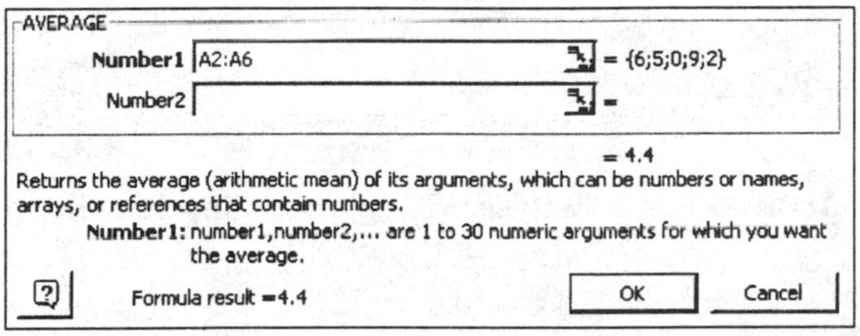

AVERAGE

Number1 A2:A6 = {6;5;0;9;2}

Number2 =

= 4.4

Returns the average (arithmetic mean) of its arguments, which can be numbers or names, arrays, or references that contain numbers.

Number1: number1,number2,... are 1 to 30 numeric arguments for which you want the average.

Formula result =4.4 OK Cancel

9. The average of the sample, 4.4, is now displayed in cell A7 of the worksheet. Repeat steps 6-9 to generate two more samples and compute the averages.

	A	B	C	D	E	F	G
1	Sample 1						
2	6						
3	5						
4	0						
5	9						
6	2						
7	22						
8	4.4						

◀

▶ Exercise 5 (pg. 23) Simulating Rolling a Six-Sided Die 60 Times

You will be simulating rolling a six-sided die 60 times and making a tally of the results.

If the PHStat add-in has not been loaded, you will need to load it before continuing. Follow the instructions in Section GS 8.2.

1. Open a new Excel worksheet and enter the digits 1 through 6 in column A. These digits represent the possible outcomes of rolling a die.

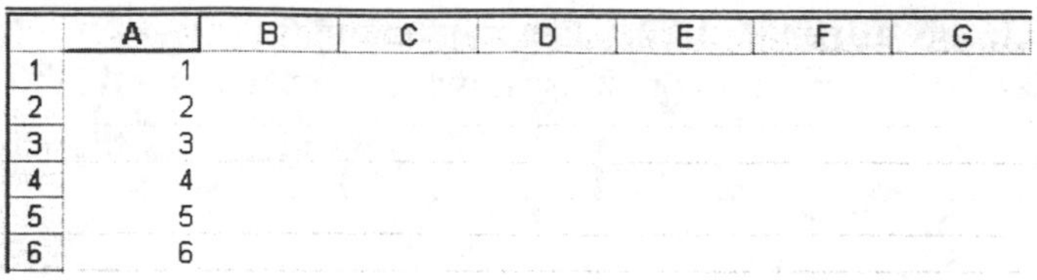

	A	B	C	D	E	F	G
1	1						
2	2						
3	3						
4	4						
5	5						
6	6						

2. At the top of the screen, click **Tools → Data Analysis**. Select **Sampling** and click **OK**.

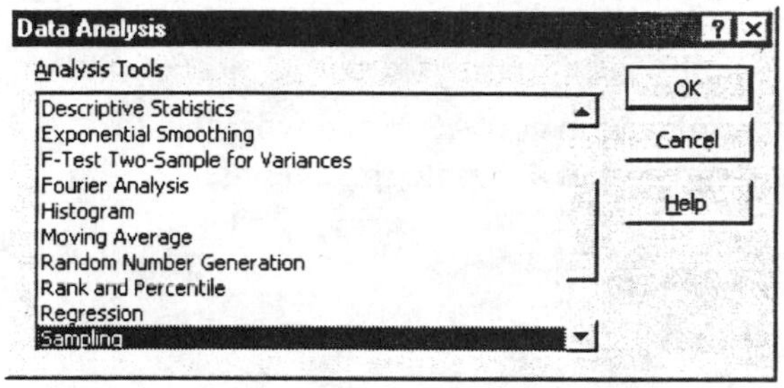

Data Analysis

Analysis Tools

Descriptive Statistics
Exponential Smoothing
F-Test Two-Sample for Variances
Fourier Analysis
Histogram
Moving Average
Random Number Generation
Rank and Percentile
Regression
Sampling

OK
Cancel
Help

3. Complete the Sampling dialog box as shown below. Click **OK**.

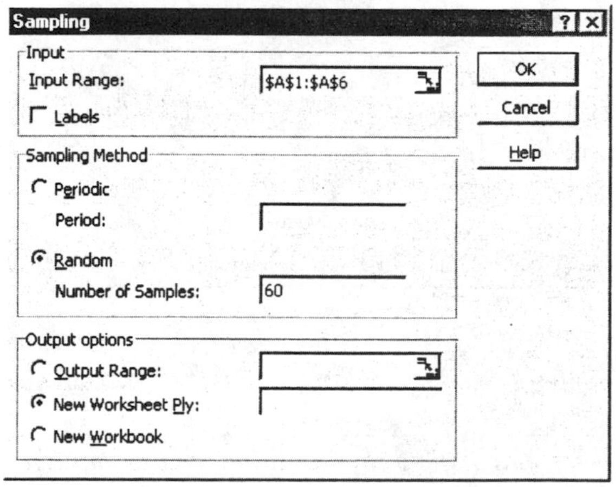

Sampling

Input
Input Range: $A$1:$A$6
Labels

Sampling Method
Periodic
Period:
Random
Number of Samples: 60

Output options
Output Range:
New Worksheet Ply:
New Workbook

OK
Cancel
Help

The output will be displayed in a new worksheet. The first 7 entries are shown below.

	A	B	C	D	E	F	G
1	1						
2	2						
3	4						
4	1						
5	3						
6	3						
7	4						

4. To make a tally of the results, first select **PHStat → One-Way Tables & Charts**. Then complete the One-Way Tables & Charts dialog box as shown below. The "Variable Cell Range" refers to the worksheet location of the output from the Random Number Generation procedure. There is no label in the first cell (A1). The output will be given the title "Rolling a Die." A bar chart will be part of the output. Click **OK**.

One-Way Tables & Charts

Data
Variable Cell Range: $A$1:$A$60
☐ First cell contains label

OK
Cancel

Output Options
Output Title: Rolling a Die
☑ Bar Chart
☐ Pie Chart
☐ Pareto Diagram

The bar chart is displayed in a sheet named "Bar Chart."

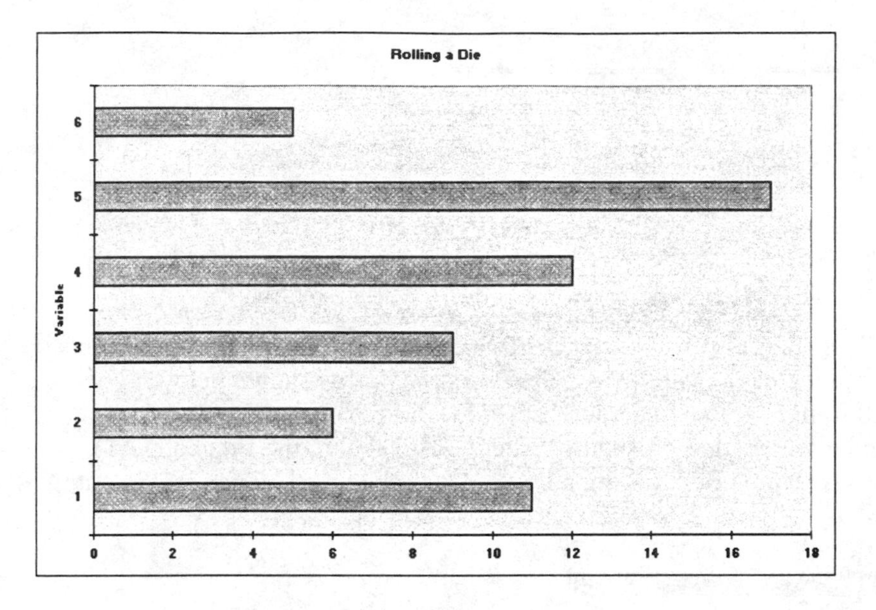

5. Click on the **One-Way Table** sheet tab at the bottom of the screen. This sheet contains a frequency distribution of the outcomes.

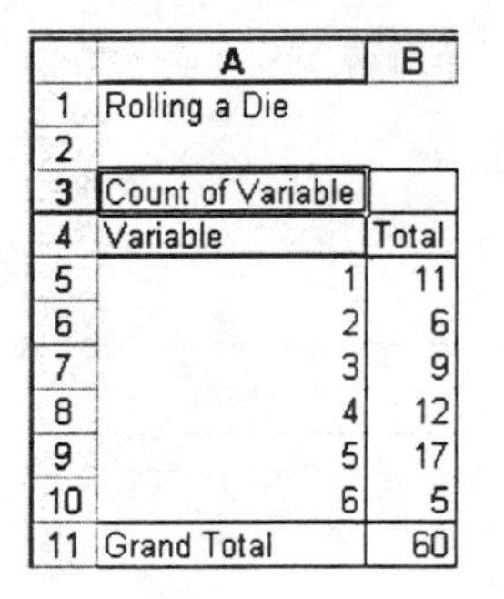

	A	B
1	Rolling a Die	
2		
3	Count of Variable	
4	Variable	Total
5	1	11
6	2	6
7	3	9
8	4	12
9	5	17
10	6	5
11	Grand Total	60

6. Click on the **DataCopy** sheet tab at the bottom of the screen. This sheet contains a copy of the generated data set with the word "Variable" inserted in cell A1.

Descriptive Statistics

CHAPTER 2

Section 2.1

► Example 7 (pg. 38)

Constructing Histograms

You will be constructing a histogram for the frequency distribution of the Internet data in Example 2 on page 33.

1. Open worksheet "internet" in the Chapter 2 folder. These data represent the number of minutes 50 Internet subscribers spent during their most recent Internet session. You will use Data Analysis located in the Tools menu to construct a histogram for these data.

   If Data Analysis does not appear as a choice in the Tools menu, you will need to load the Microsoft Excel Analysis ToolPak add-in. Follow the procedure in Section GS 8.1 before continuing.

   ---

2. Excel's histogram procedure uses grouped data to generate a frequency distribution and a frequency histogram. The procedure requires that you indicate a "bin" for each class. The number that you specify for each bin is actually the upper limit of the class. The upper limits for the Internet data are given on page 31 of your text and are based on a class width of 12. You see that 18 is the upper limit for the first class, 30 is the upper limit for the second class, and so on. The bin for the first class will contain a count of all observations less than or equal to 18, the bin for the

second class will contain a count of all observations between 19 and 30, and so on. Enter **Bin** in cell C1 and key in the upper limits in column C as shown below.

	A	B	C	D	E	F	G
1	Internet subscribers		Bin				
2	50		18				
3	40		30				
4	41		42				
5	17		54				
6	11		66				
7	7		78				
8	22		90				

3. Click **Tools** → **Data Analysis**. Select **Histogram**, and click **OK**.

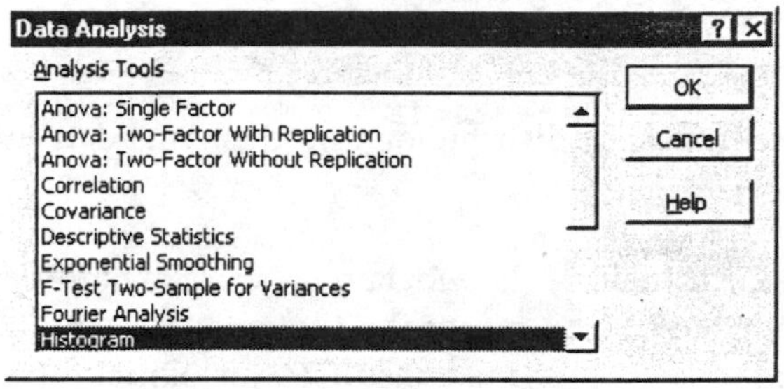

Data Analysis

Analysis Tools

Anova: Single Factor
Anova: Two-Factor With Replication
Anova: Two-Factor Without Replication
Correlation
Covariance
Descriptive Statistics
Exponential Smoothing
F-Test Two-Sample for Variances
Fourier Analysis
Histogram

OK
Cancel
Help

4. Complete the Histogram dialog box as shown below and click **OK**.

To enter the input and bin ranges quickly, follow these steps. First click in the Input Range field of the dialog box. Then, in the worksheet, click and drag from cell A1 to cell A51. You will then see $A$1:$A$51 displayed in the Input Range field. Next click in the Bin Range field. Then click and drag from cell C1 through cell C8 in the worksheet. You will see $C$1:$C$8 displayed in the Bin Range field.

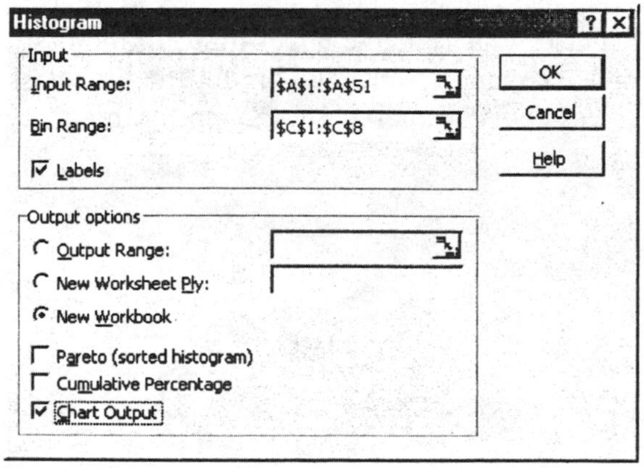

Histogram

Input
Input Range: $A$1:$A$51
Bin Range: $C$1:$C$8
☑ Labels

Output options
○ Output Range:
○ New Worksheet Ply:
◉ New Workbook

☐ Pareto (sorted histogram)
☐ Cumulative Percentage
☑ Chart Output

OK
Cancel
Help

Note the checkmark in the box to the left of Labels. In your worksheet, "Internet subscribers" appears in cell A1, and "Bin" appears in cell C1. Because you included these cells in the Input Range and Bin Range, respectively, you need to let Excel know that these cells contain labels rather than data. Otherwise Excel will attempt to use the information in these cells when constructing the frequency distribution and histogram.

You should see output similar to the output displayed below.

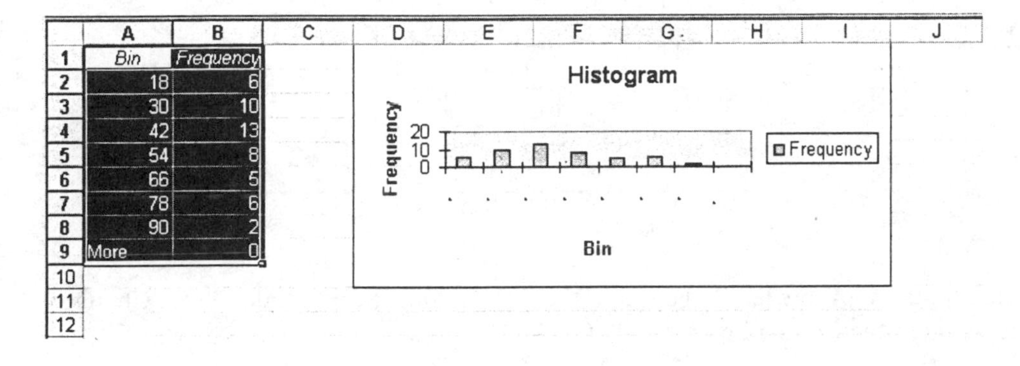

Bin	Frequency
18	6
30	10
42	13
54	8
66	5
78	6
90	2
More	0

If you want to change the height and width of the chart, begin by clicking anywhere within the figure. Handles appear along the perimeter. You can change the shape of the figure by clicking on a handle and dragging it. You can make the figure small and short or you can make it tall and wide. You can also click within the figure and drag it to move it to a different location on the worksheet.

---

5. You will now follow steps to modify the histogram so that it is displayed in a more informative and attractive manner. First, make the chart taller so that it is easier to read. To do this, first click within the figure near a border. Black square handles appear. Click on the center handle on the bottom border of the figure and drag it down a few rows.

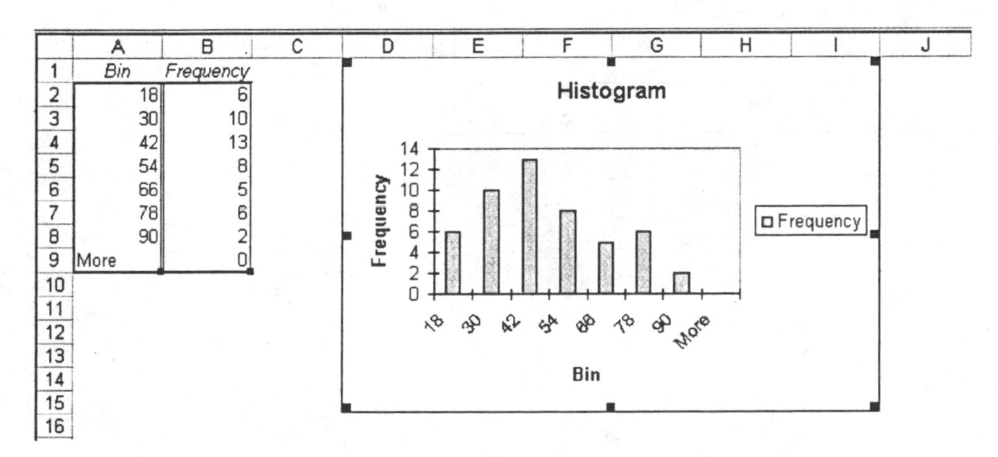

Bin	Frequency
18	6
30	10
42	13
54	8
66	5
78	6
90	2
More	0

6. Next remove the space between the vertical bars. **Right click** on one of the vertical bars. Select **Format Data Series** from the shortcut menu that appears.

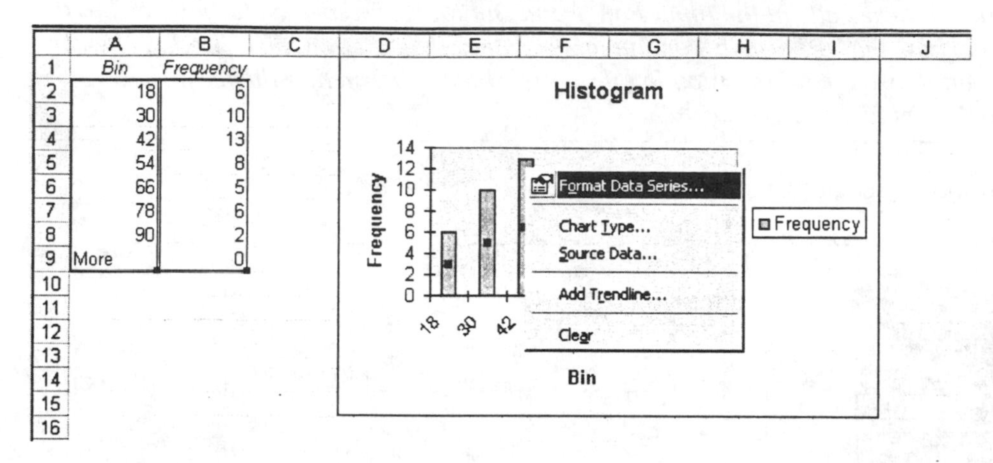

7. Click on the **Options** tab at the top of the Format Data Series dialog box. Change the value in the Gap width box to 0. Click **OK**.

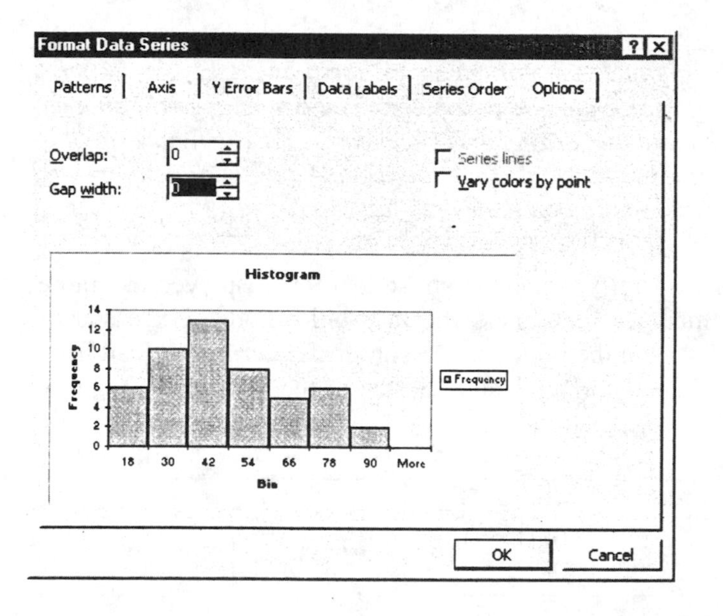

8. Change the X-axis values from upper limits to midpoints. The midpoints are displayed in a table on page 33 of your textbook. Enter these midpoints in column C of the Excel worksheet as shown below.

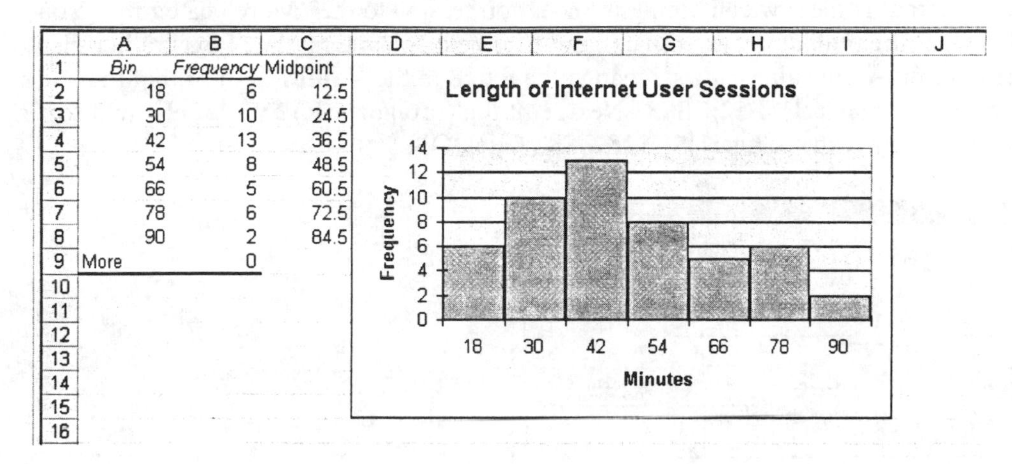

9. **Right click** on a vertical bar. Select **Source Data** from the shortcut menu that appears.

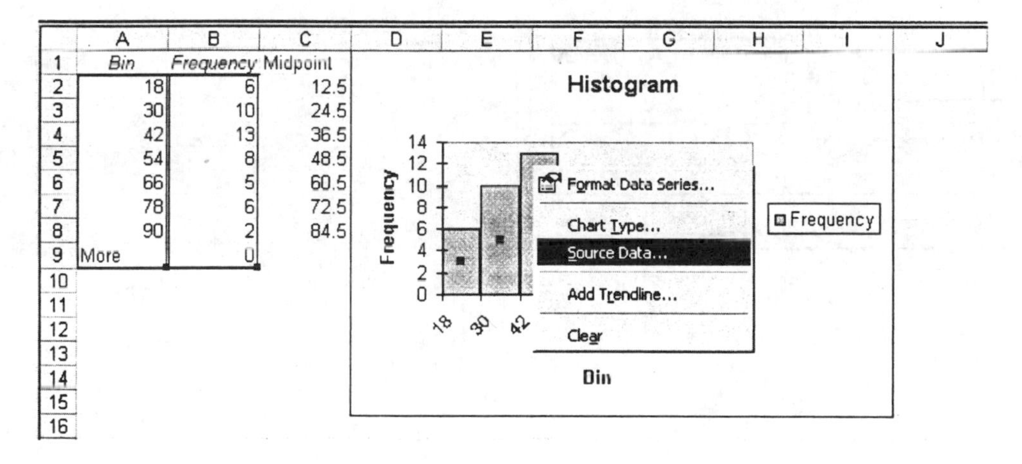

10. Click on the **Series** tab at the top of the Source Data dialog box. The ranges displayed in the Values field and the Category (X) labels field refer to the frequency distribution table in the top left of the worksheet. You do not want to include row 9, because that is the row containing information related to the "More" category. You also want the midpoint values in column C to be displayed on the X axis rather than the column A bin values. First, change the 9 to 8 in the **Values** field so that the entry reads **=Sheet1!$B$2:$B$8**. Next, edit the **Category (X) axis labels** field so that the entry reads **=Sheet1!$C$2:$C$8**. Click **OK**.

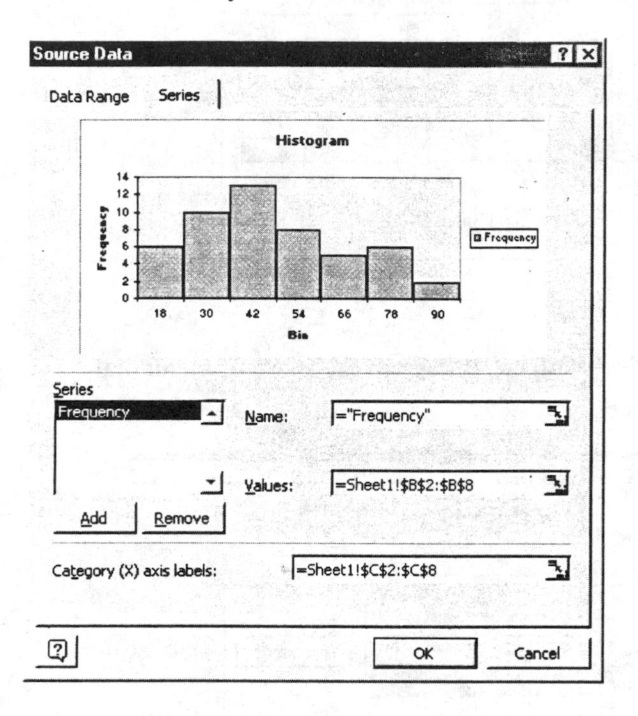

11. You will use Chart Options to modify three aspects of the histogram: Titles, gridlines, and legend. Right click in the gray plot area of the chart and select **Chart Options** from the shortcut menu that appears.

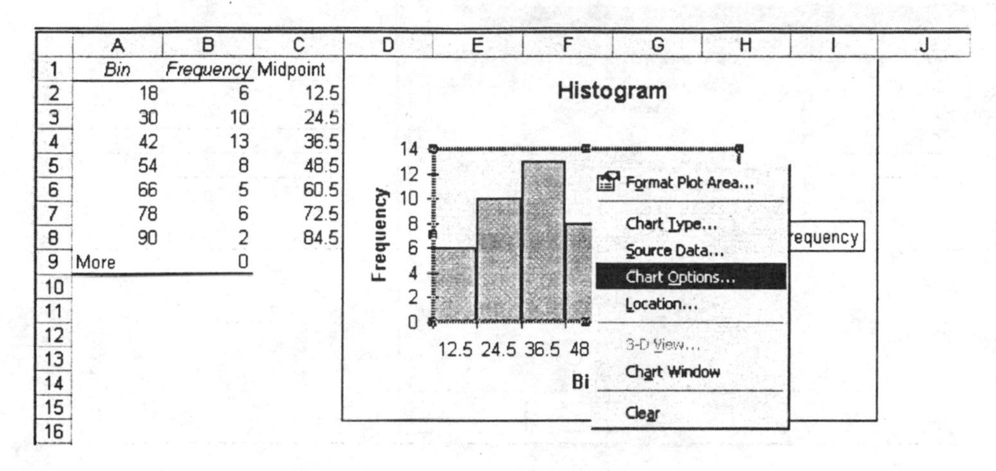

12. Click on the **Titles** tab at the top of the Chart Options dialog box. Change the Chart title from "Histogram" to **Internet Usage**. Change the Category (X) axis label from "Bin" to **Time online (in minutes)**.

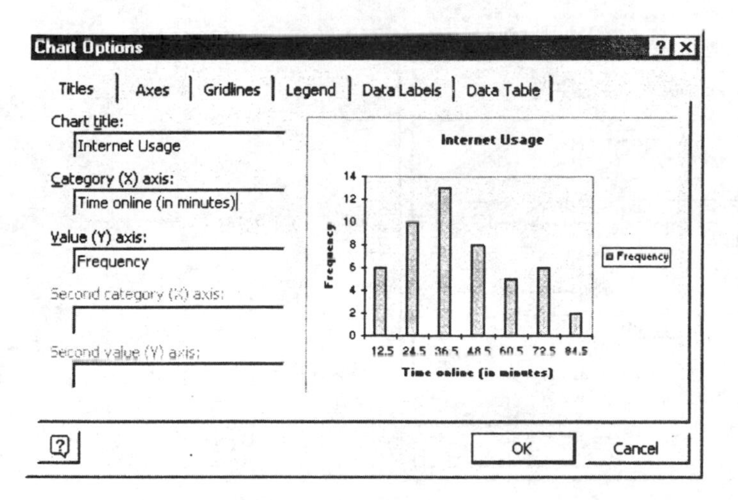

13. Click on the **Gridlines** tab at the top of the Chart Options dialog box. Under Value (Y) axis, click in the **Major gridlines** box so that a checkmark appears there.

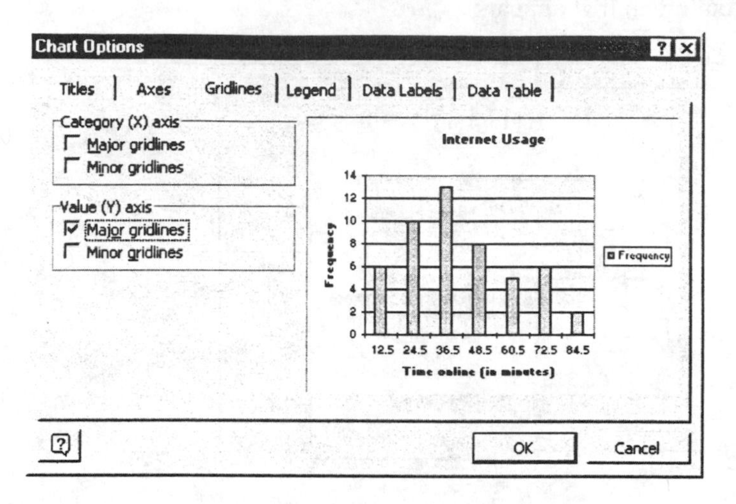

14. Click on the **Legend** tab. To remove the frequency legend displayed at the right of the histogram chart, click in the **Show legend** box to remove the checkmark. Click **OK**. Your histogram chart should now appear similar to the one displayed below.

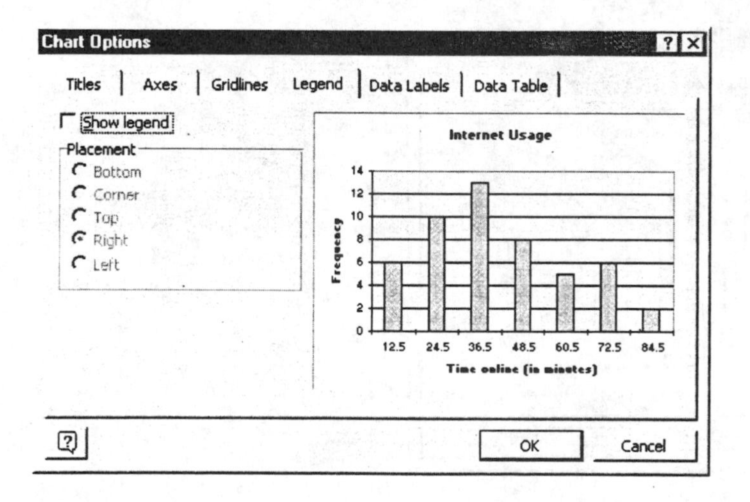

▶ Exercise 17 (pg. 41) Constructing a Frequency Distribution and Frequency Histogram for the July Sales Data

1. Open worksheet "ex2_1-17" in the Chapter 2 folder.

2. Sort the data so that it is easy to identify the minimum and maximum data entries. In the sorted data set, you see that the minimum sales value is 1000 and that the maximum is 7119.

For instructions on how to sort, see Sections GS 5.1 and GS 5.2.

3. Calculate class width using the formula given in your textbook:

$$\text{Class width} = \frac{\text{Maximum data entry} - \text{Minimum data entry}}{\text{Number of classes}}$$

The exercise instructs you to use 6 for the number of classes.

$$\text{Class width} = \frac{7119 - 1000}{6} = 1019.83\text{. Round up to 1020.}$$

4. The textbook instructs you to use the minimum data entry as the lower limit of the first class. To find the remaining lower limits, add the class width of 1020 to the lower limit of each previous class. Enter **Lower Limit** in cell B1 of the worksheet, and enter **1000** in cell B2. You will now use a formula to compute the remaining lower limits, and you will have Excel do these calculations for you. In cell B3, enter the formula **=B2+1020** as shown below. Press **[Enter]**.

	A	B	C	D	E	F	G
1	July Sales	Lower Limit					
2	1000	1000					
3	1030	=B2+1020					

5. Click on cell **B3** where 2020 now appears and copy the formula in cell B3 to cells B4 through B8. Because 7119 is the maximum data entry, you don't need 7120 for the histogram. However, you will be using 7120 when calculating the upper limit for the last class interval.

	A	B	C	D	E	F	G
1	July Sales	Lower Limit					
2	1000	1000					
3	1030	2020					
4	1077	3040					
5	1355	4060					
6	1500	5080					
7	1512	6100					
8	1643	7120					

6. Enter **Upper Limit** in cell C1 of the worksheet. The textbook says that the upper limit is equal to one less than the lower limit of the next higher class. You will use a formula to compute the upper limits, and you will have Excel do the computations for you. Click in cell C2 and enter the formula **=B3-1** as shown below. Press **[Enter]**.

	A	B	C	D	E	F	G
1	July Sales	Lower Limit	Upper Limit				
2	1000	1000	=B3-1				
3	1030	2020					

7. Copy the formula in cell C2 (where 2019 now appears) to cells C3 through C7. These are the upper limits that you will use as bins for constructing the histogram chart.

	A	B	C	D	E	F	G
1	July Sales	Lower Limit	Upper Limit				
2	1000	1000	2019				
3	1030	2020	3039				
4	1077	3040	4059				
5	1355	4060	5079				
6	1500	5080	6099				
7	1512	6100	7119				
8	1643	7120					

8. Click **Tools → Data Analysis**. Select **Histogram** and click **OK**.

If Data Analysis does not appear as a choice in the Tools menu, you will need to load the Microsoft Excel Analysis ToolPak add-in. Follow the procedure in Section GS 8.1 before continuing

---

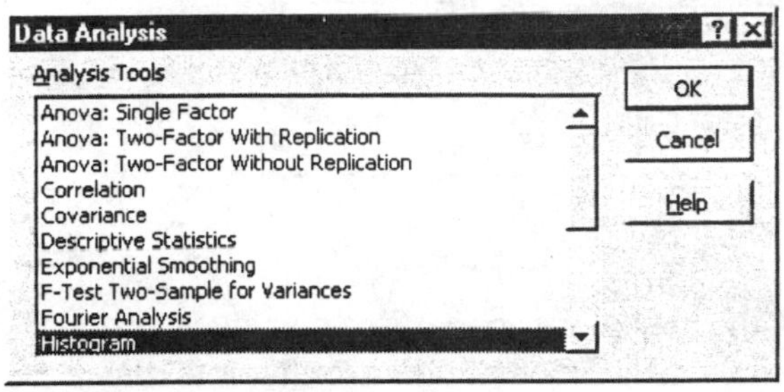

Data Analysis
Analysis Tools
Anova: Single Factor
Anova: Two-Factor With Replication
Anova: Two-Factor Without Replication
Correlation
Covariance
Descriptive Statistics
Exponential Smoothing
F-Test Two-Sample for Variances
Fourier Analysis
Histogram
OK
Cancel
Help

9. Complete the fields in the Histogram dialog box as shown below. The July sales data are located in cells A1 through A23 of the worksheet. The bins (upper limits) are located in cells C1 through C7. The top cells in these ranges are labels—"July Sales" and "Upper Limit." The output will be placed in a new Excel workbook. The output will include a chart. Click **OK**.

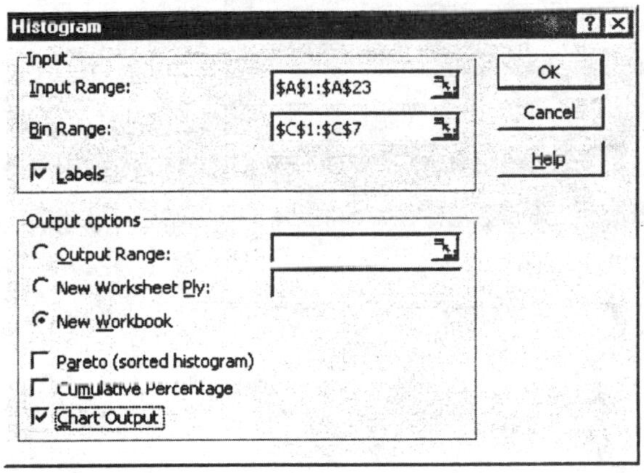

Histogram
Input
Input Range: $A$1:$A$23
Bin Range: $C$1:$C$7
Labels
OK
Cancel
Help
Output options
Output Range:
New Worksheet Ply:
New Workbook
Pareto (sorted histogram)
Cumulative Percentage
Chart Output

You should see output similar to the output displayed below.

	A	B
1	Upper Limi	Frequency
2	2019	12
3	3039	3
4	4059	2
5	5079	3
6	6099	1
7	7119	1
8	More	0

Histogram

Frequency 20 10 0

Frequency

Upper Limit

10. You will now follow steps to modify the histogram so that it is displayed in a more accurate and informative manner. First, make the chart taller so that it is easier to read. To do this, first click within the figure near a border. Black square handles appear. Click on the center handle on the bottom border of the figure and drag it down a few rows.

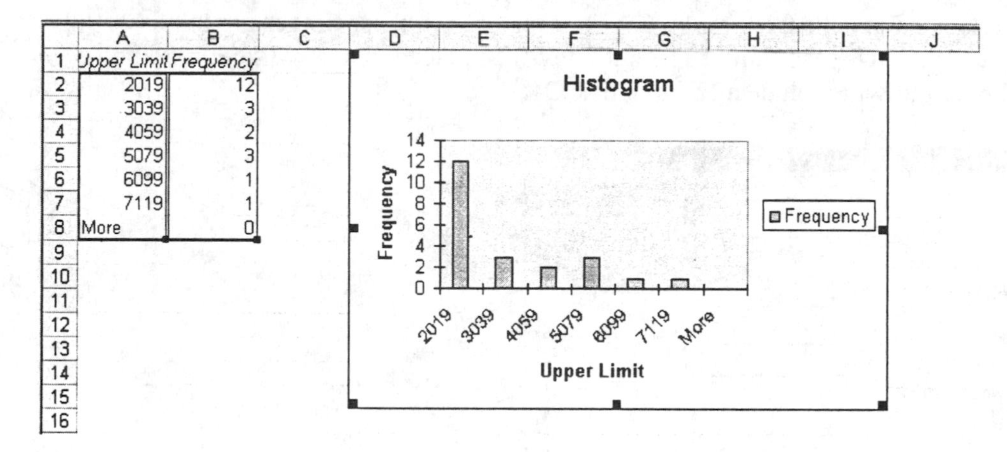

11. Remove the space between the vertical bars. **Right** click on one of the vertical bars. Select **Format Data Series** from the shortcut menu that appears.

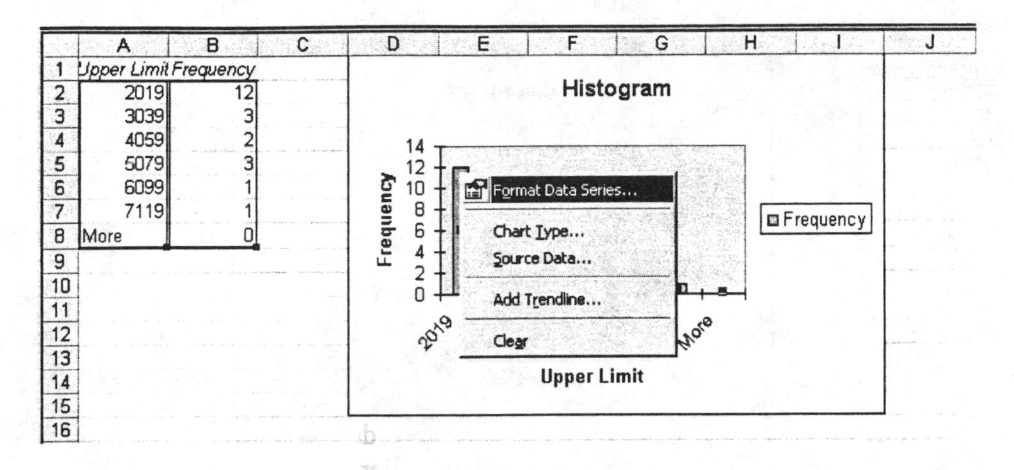

12. Click on the **Options** tab at the top of the Format Data Series dialog box. Change the value in the Gap width box to 0. Click **OK**.

Format Data Series

Patterns | Axis | Y Error Bars | Data Labels | Series Order | Options

Overlap: 0

Gap width: 0

Series lines

Vary colors by point

Histogram

Frequency

2019 3039 4059 5079 6099 7119 More

Upper Limit

OK Cancel

13. Delete the word "More" from the X axis. **Right click** on a vertical bar. Select **Source Data** from the shortcut menu that appears.

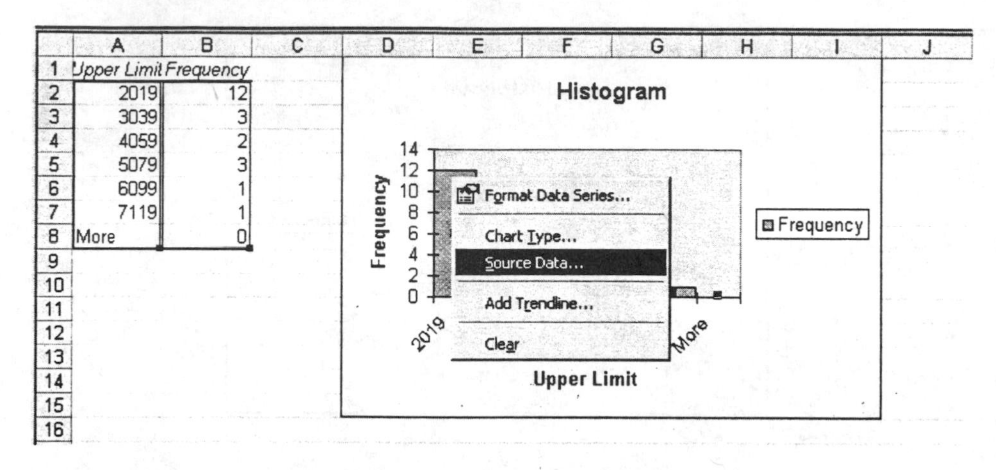

14. Click on the **Series** tab at the top of the Source Data dialog box. You do not want to include row 8 of the frequency distribution because that is the row containing information related to the "More" category. Change the 8 to 7 in the **Values** field so that it reads **=Sheet1!$B$2:$B$7**. Change the 8 to 7 in the **Category (X) axis labels** field so that it reads **=Sheet1!$A$2:$A$7**. Click **OK**.

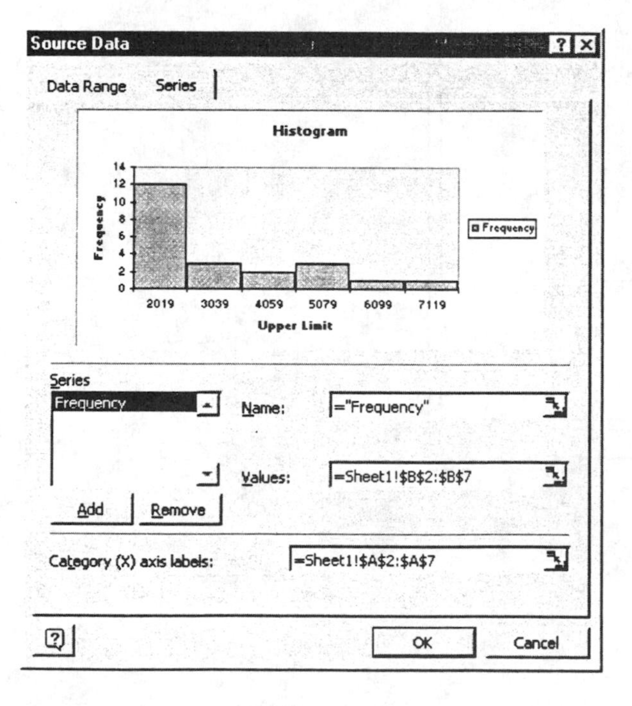

15. You will use Chart Options to modify three aspects of the histogram: Titles, gridlines, and legend. **Right click** in the gray plot area of the chart and select **Chart Options** from the shortcut menu that appears.

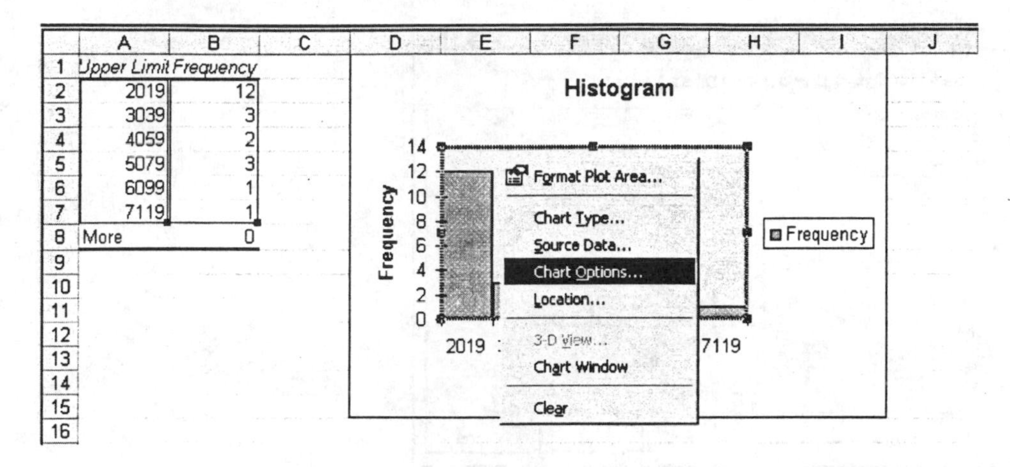

16. Click on the **Titles** tab at the top of the Chart Options dialog box. Change the Chart title from "Histogram" to **July Sales**. Change the Category (X) label from "Upper Limit" to **Dollars**.

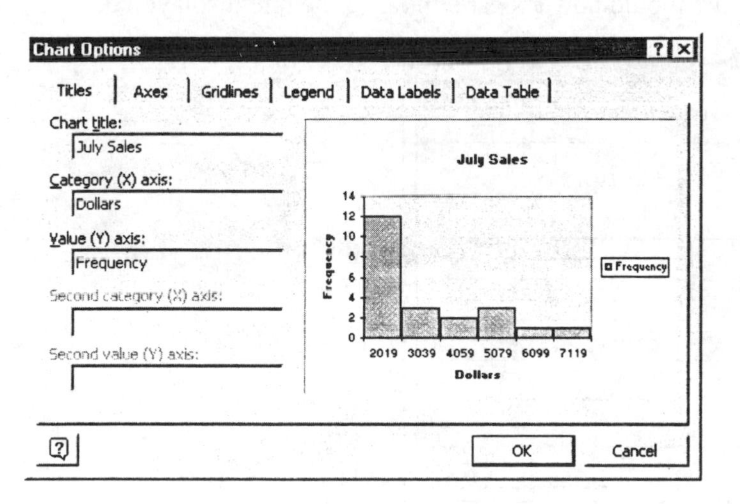

17. Click on the **Gridlines** tab at the top of the Chart Options dialog box. Under Value (Y) axis, click in the **Major gridlines** box.

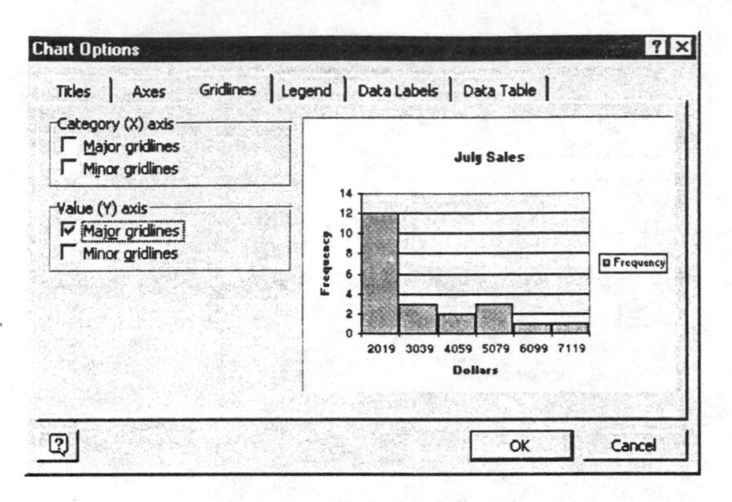

18. Click on the **Legend** tab. To remove the frequency legend displayed at the right of the histogram chart, click in the **Show legend** box to remove the checkmark. Click **OK**. Your histogram chart should now appear similar to the one displayed below.

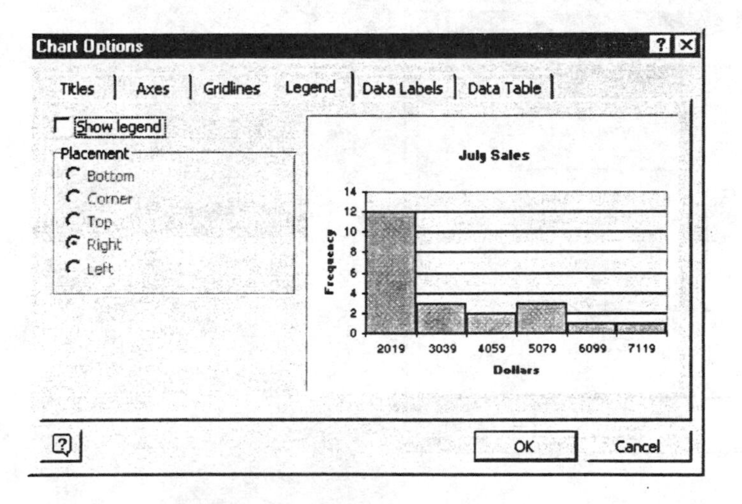

Section 2.2

► Example 1 (pg. 44)

Constructing a Stem-and-Leaf Plot

If the PHStat add-in has not been loaded, you will need to load it before continuing. Follow the instructions in Section GS 8.2.

1. Open worksheet "AL_RBIs" in the Chapter 2 folder.

2. At the top of the screen, click **PHStat** and select **Stem-and-Leaf Display**.

3. Complete the entries in the Stem-and-Leaf Display dialog box as shown below. Click **OK**.

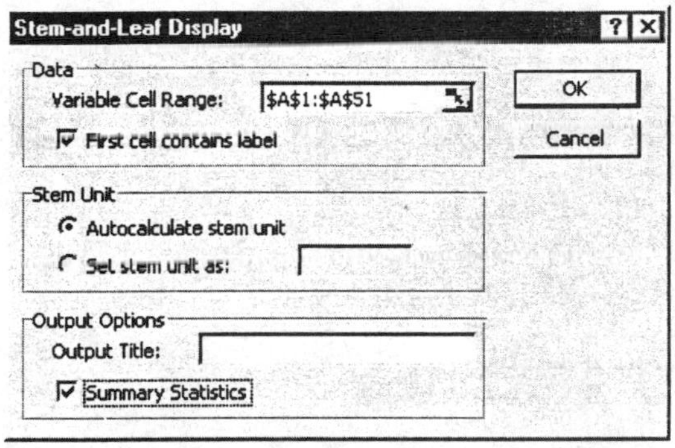

Stem-and-Leaf Display
Data
Variable Cell Range: $A$1:$A$51
First cell contains label
OK
Cancel
Stem Unit
Autocalculate stem unit
Set stem unit as:
Output Options
Output Title:
Summary Statistics

The output is displayed in a new sheet named "StemLeafPlot."

	A	B	C	D	E	F	G
1				Stem-and-Leaf Display			
2				for A.L. RBIs			
3				Stem unit: 10			
4							
5	Statistics			7 | 8			
6	Sample Size	50		8 |			
7	Mean	125.14		9 |			
8	Median	123.5		10 | 5 8 9 9 9			
9	Std. Deviation	14.53919		11 | 2 2 2 3 4 6 7 8 8 8 9 9 9			
10	Minimum	78		12 | 1 1 2 2 2 3 4 4 6 6 6 6 6 9 9			
11	Maximum	159		13 | 0 2 3 3 4 9 9			
12				14 | 0 2 4 5 5 7 8			
13				15 | 5 9			
14							

◄

► Example 4 (pg. 47) Constructing a Pie Chart

1. Open a new, blank Excel worksheet and enter the transportation frequency table as shown below.

	A	B	C	D	E	F	G
1	Transportation	Passengers					
2	Bus	359					
3	Air	499.1					
4	Subway	351.6					
5	Amtrak	19.7					

2. Click on any cell within the table. Then click **Insert → Chart.**

If a cell within the table is activated when you select Chart, the data range will automatically be entered into the Chart dialog box.

3. Under Chart type, in the Chart Type dialog box, select **Pie**. Under Chart sub-type, select the first one in the top row by clicking on it. Click **Next** at the bottom of the dialog box.

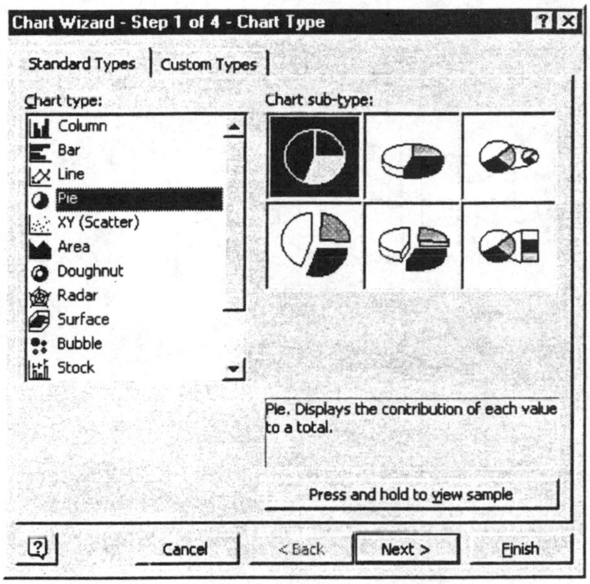

4. Check the accuracy of the data range in the Chart Source Data dialog box. It should read **=Sheet1!$A$1:$B$5**. Make any necessary corrections. Then click **Next**.

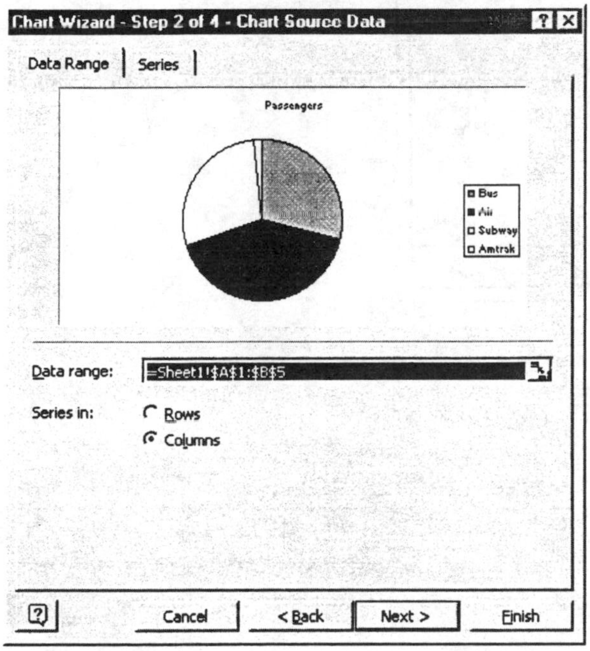

5. Click the **Titles** tab at the top of the Chart Options dialog box. Change the title from "Passengers" to **Intercity Passenger Travel**.

Chart Wizard - Step 3 of 4 - Chart Options

Titles | Legend | Data Labels

Chart title:

Intercity Passenger Travel

Category (X) axis:

Value (Y) axis:

Second category (X) axis:

Second value (Y) axis:

Intercity Passenger Travel

Bus
Air
Subway
Amtrak

Cancel | < Back | Next > | Finish

6. Click the **Data Labels** tab at the top of the Chart Options dialog box. Click the button next to **Show label and percent** so that a black dot appears there. Click **Next**.

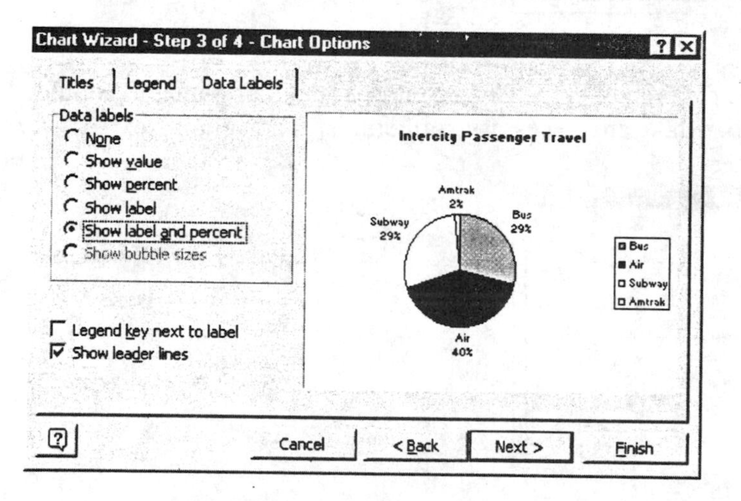

7. The Chart Location dialog box presents two options for placing your chart output. For this example, select **As object in**. Click **Finish**.

8. Your textbook shows one decimal place in the percent values, whereas the chart generated by Excel shows no decimal places. To change the number of decimal places, first **right click** directly on one of the percent values. Then click on **Format Data Labels** in the shortcut menu that appears.

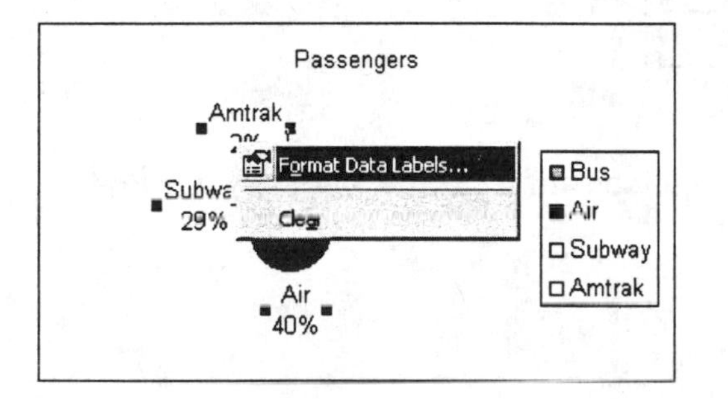

9. Click the **Number** tab at the top of the Format Data Labels dialog box. Click the up arrow to the right of the decimal places field so that **1** is displayed in the field. Click **OK**.

Format Data Labels
Patterns | Font | Number | Alignment
Category:
General
Number
Currency
Accounting
Date
Time
Percentage
Fraction
Scientific
Text
Special
Custom
Sample
29.2%
Decimal places: 1
Linked to source
Percentage formats multiply the cell value by 100 and displays the result with a percent symbol.
OK
Cancel

Your pie chart for the transportation data should look similar to the one shown below.

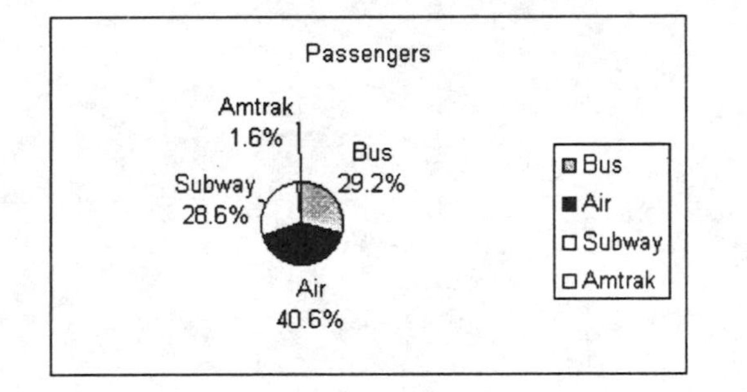

► Example 5 (pg. 48) Constructing a Pareto Chart

1. Open a new, blank Excel worksheet, and enter the inventory shrinkage information as shown below.

	A	B	C	D	E	F	G
1	Cause	Millions of dollars					
2	Administrative error	7.8					
3	Employee theft	15.6					
4	Shoplifting	14.7					
5	Vendor fraud	2.9					

2. In a Pareto chart, the vertical bars are placed in order of decreasing height. Excel's Chart function will display the information in the order in which it appears in the worksheet. So, you need to sort the information in descending order by Millions of dollars.

For instructions on how to sort, see Sections GS 5.1 and GS 5.2.

	A	B	C	D	E	F	G
1	Cause	Millions of dollars					
2	Employee theft	15.6					
3	Shoplifting	14.7					
4	Administrative error	7.8					
5	Vendor fraud	2.9					

3. After sorting the data, click on any cell within the data table. Then click **Insert → Chart**.

4. In the Chart Type dialog box, select a **Column** chart type. Click on the top left chart sub-type to select it. Click **Next**.

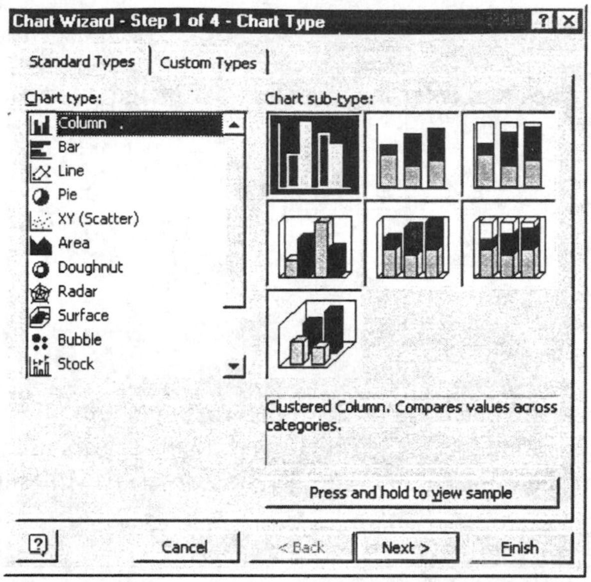

Chart Wizard - Step 1 of 4 - Chart Type

Standard Types | Custom Types

Chart type: Column, Bar, Line, Pie, XY (Scatter), Area, Doughnut, Radar, Surface, Bubble, Stock

Chart sub-type:

Clustered Column. Compares values across categories.

Press and hold to view sample

Cancel | < Back | Next > | Finish

5. Check the range displayed in the Chart Source Data dialog box to make sure it is accurate. It should read =Sheet1!$A1:$B$5. Make any necessary corrections. Then click **Next**.

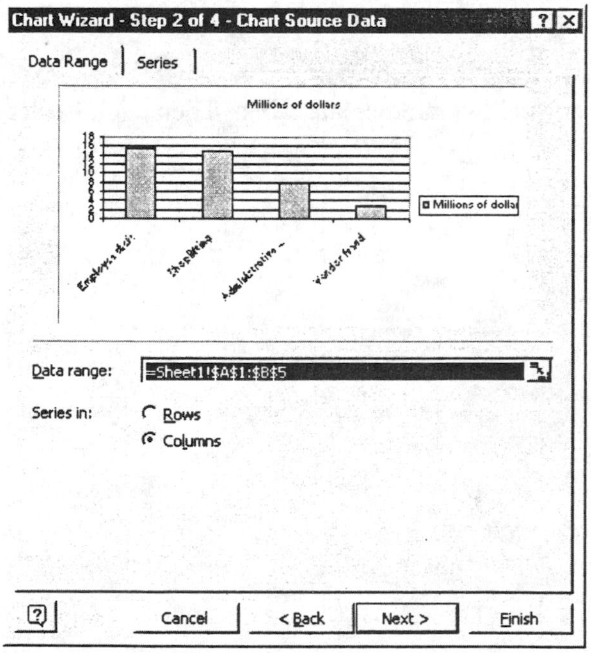

Chart Wizard - Step 2 of 4 - Chart Source Data

Data Range | Series

Millions of dollars

Data range: =Sheet1!$A$1:$B$5

Series in: Rows / Columns

Cancel | < Back | Next > | Finish

6. Click the **Titles** tab at the top of the Chart Options dialog box. In the Chart title field, enter **Causes of Inventory Shrinkage**. In the Category (X) axis field, enter **Cause**. In the Value (Y) axis field, enter **Millions of dollars**.

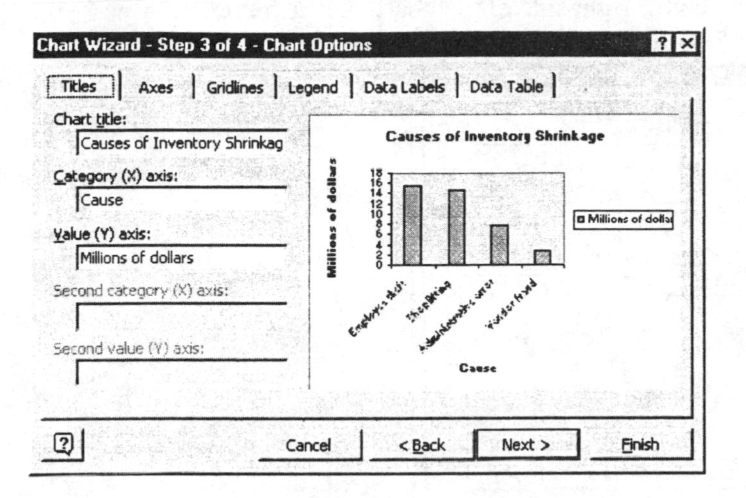

7. Click the **Gridlines** tab at the top of the Chart Options dialog box. Below Value (Y) axis, click in the box to the left of **Major gridlines** so that a checkmark appears there.

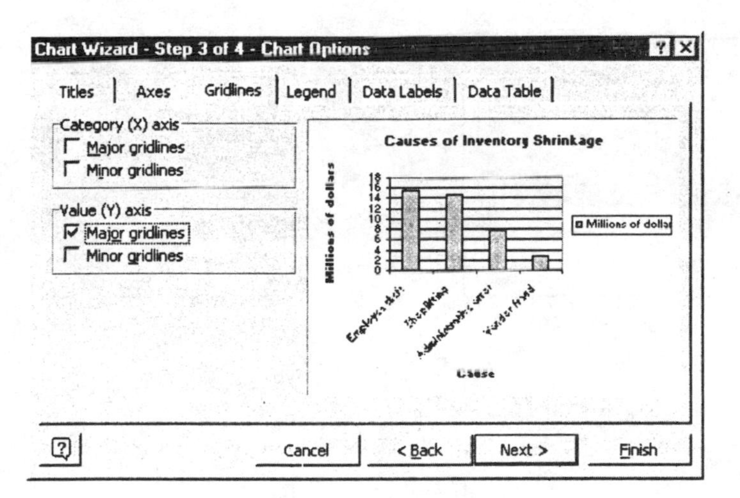

8. Click the **Legend** tab at the top of the Chart Options dialog box. Click in the box to the left of **Show legend** so that no checkmark appears there. This will remove the Millions of dollars legend on the right side of the chart. Click **Next**.

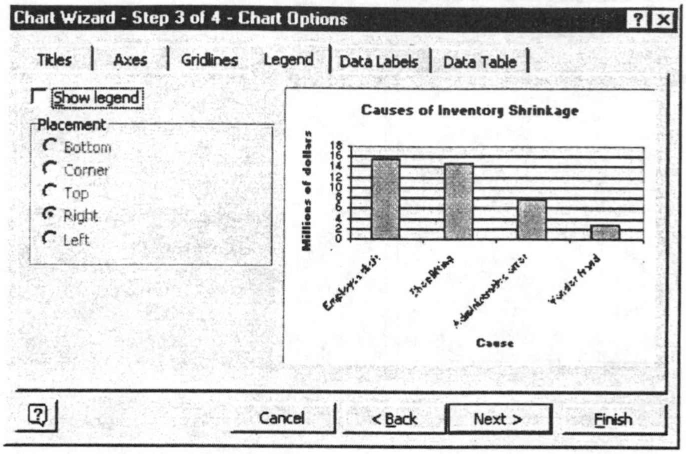

Chart Wizard - Step 3 of 4 - Chart Options

Titles | Axes | Gridlines | Legend | Data Labels | Data Table

Show legend

Placement

Bottom

Corner

Top

Right

Left

Causes of Inventory Shrinkage

Millions of dollars

18 16 14 12 10 8 6 4 2 0

Cause

Cancel | < Back | Next > | Finish

9. The Chart Location dialog box presents two options for placement of the chart. For this example, select **As object in**. Click **Finish**.

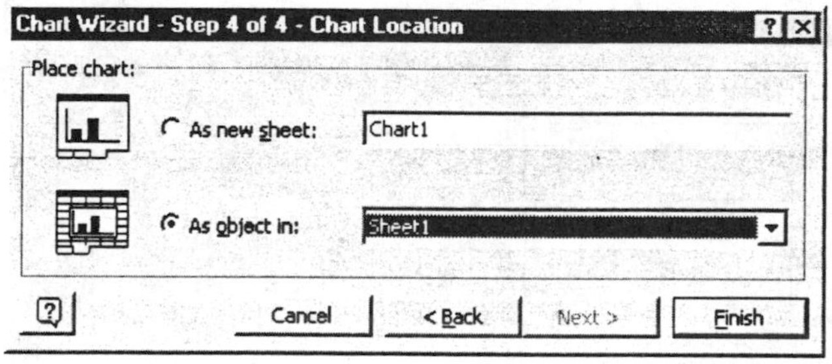

Chart Wizard - Step 4 of 4 - Chart Location

Place chart:

As new sheet: Chart1

As object in: Sheet1

Cancel | < Back | Next > | Finish

10. Make the chart taller so that it is easier to read. To do this, first click within the figure near a border so that black square handles appear. Click on the center handle at the bottom border of the figure and drag it down a few rows. Your chart should look similar to the one shown below.

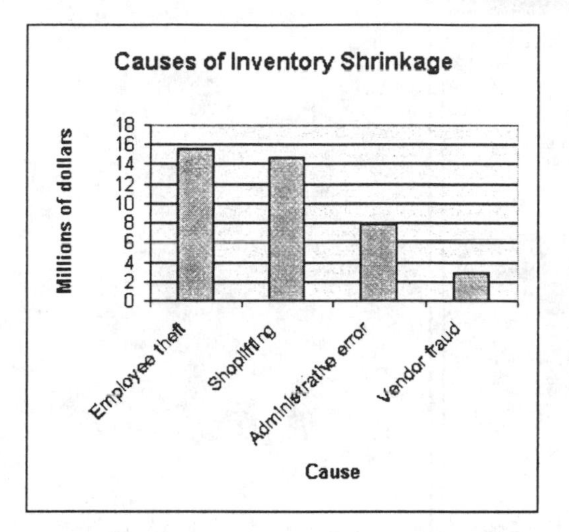

11. Remove the space between the vertical bars. **Right click** on one of the vertical bars. Select **Format Data Series** from the shortcut menu that appears.

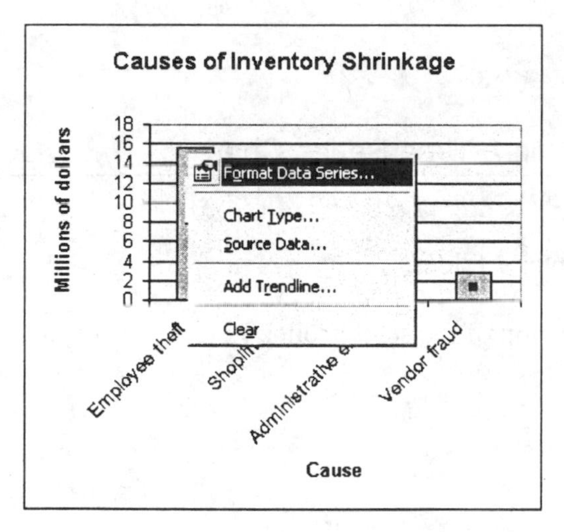

12. Click the **Options** tab at the top of the Format Data Series dialog box. Change the value in the Gap width box to 0. Click **OK**. Your histogram chart should now look similar to the one displayed below.

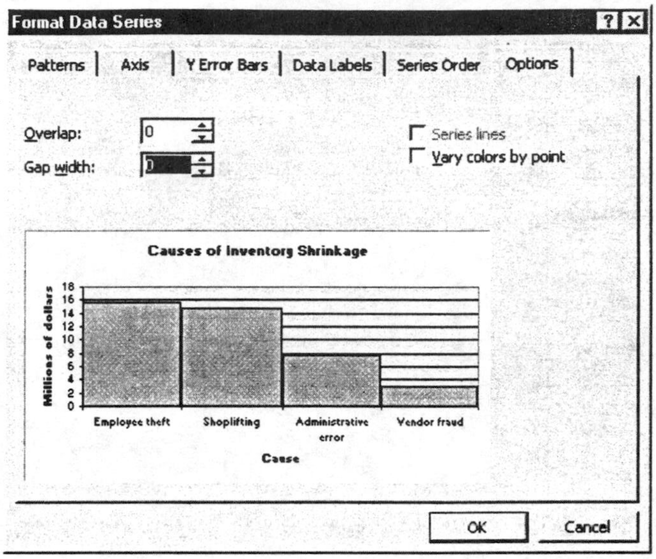

▶ Example 7 (pg. 50)

Constructing a Time Series Chart

1. Open worksheet "cellphone" in the Chapter 2 folder.

2. Click on any cell within the table. At the top of the screen, click **Insert** and select **Chart** from the menu that appears.

3. In the Chart Type dialog box, select the **XY (Scatter)** chart type by clicking on it. Then click on the topmost chart sub-type. Click **Next**.

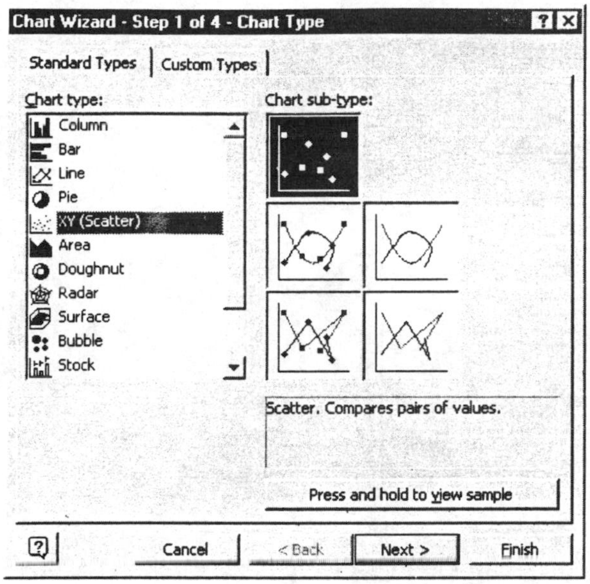

4. In the Chart Source Data dialog box, edit the data range so that it does not include the average bill data in column C of the worksheet. The entry in the data range field should be **=cellphone!$A$1:$B$11**. Click **Next**.

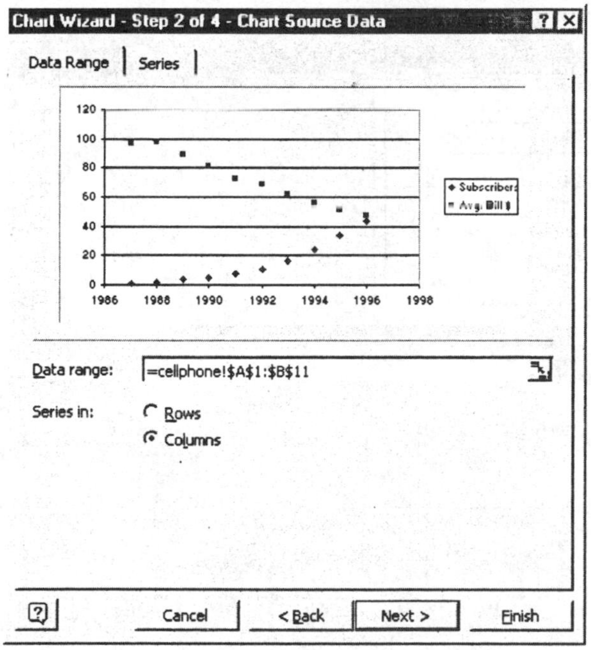

5. Click the **Titles** tab at the top of the Chart Options dialog box. In the Chart title field, change the title to **Cellular Telephone Subscribers by Year**. In the Value (X) Axis field, enter **Year**. In the Value (Y) axis field, enter **Subscribers (in millions)**.

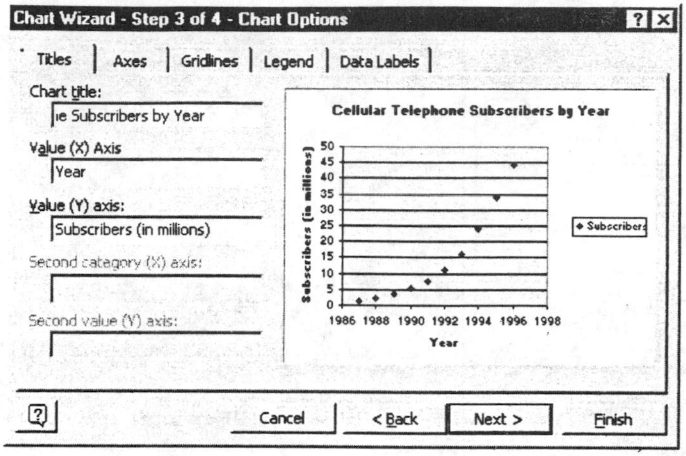

6. Click the **Legend** tab at the top of the Chart Options dialog box. Click in the box to the left of **Show Legend** so that a checkmark does not appear there. This removes the Subscribers legend from the right side of the scatter plot. Click **Next**.

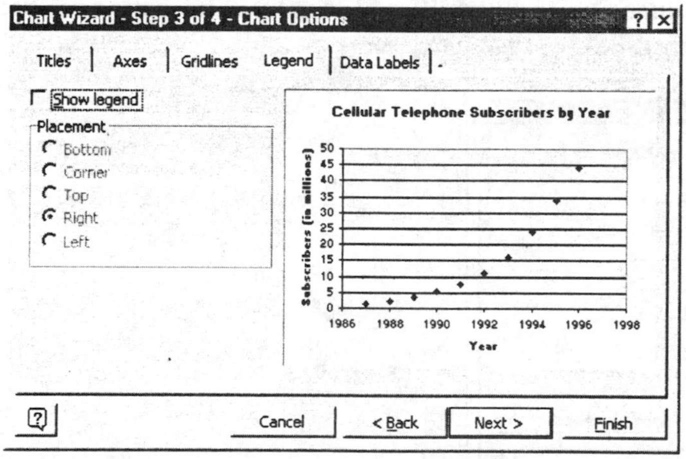

7. The Chart Location dialog box presents two options for placing the scatter plot. For this example, select **As object in** the same worksheet as the data. To do this, click the button to the left of **As object in** so that a black dot appears there. Click **Finish**.

Chart Wizard - Step 4 of 4 - Chart Location

Place chart:

As new sheet: Chart1

As object in: cellphone

Cancel | < Back | Next > | Finish

8. The scatter plot that you will obtain is quite compressed. Make it taller so that it is easier to read. To do this, first click within the figure near a border. Black square handles appear. Then click on the center handle on the bottom border of the figure and drag it down a few rows.

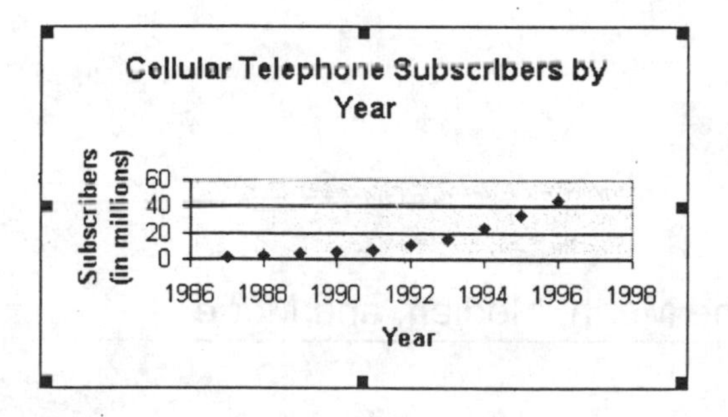

Your scatter plot should now look similar to the one shown below.

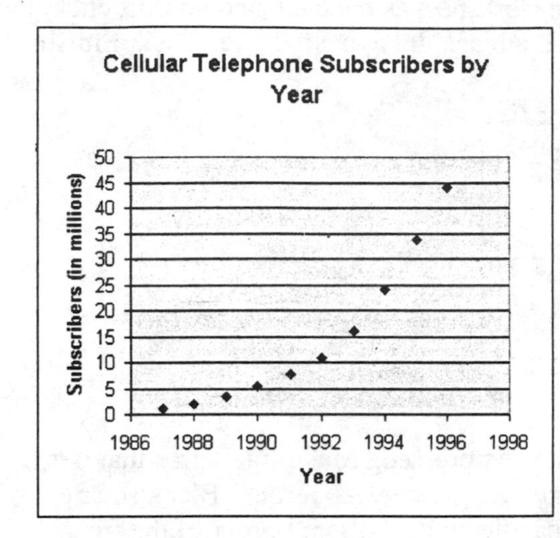

◀

Section 2.3

► Example 6 (pg. 58) — Finding the Mean, Median, and Mode

If the PHStat add-in has not been loaded, you will need to load it before continuing. Follow the instructions in Section GS 8.2.

---

1. Open worksheet "ages" in the Chapter 2 folder.

2. At the top of the screen, click **Tools → Data Analysis**. Select **Descriptive Statistics** and click **OK**.

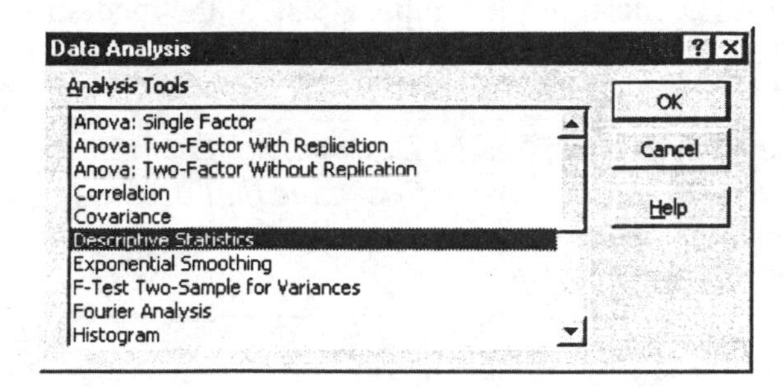

3. Complete the Descriptive Statistics dialog box as shown below. The entry in the **Input Range** field presents the worksheet location of the age data. The checkmark to the left of **Labels in First Row** lets Excel know that the entry in cell A1 is a label and is not to be included in the computations. The output will be placed in a **New Worksheet.** The checkmark to the left of **Summary statistics** requests that the output include summary statistics for the specified set of data. Click **OK**.

Descriptive Statistics
Input
Input Range: $A$1:$A$22
Grouped By: Columns
Rows
Labels in First Row
Output options
Output Range:
New Worksheet Ply:
New Workbook
Summary statistics
Confidence Level for Mean: 95 %
Kth Largest: 1
Kth Smallest: 1
OK
Cancel
Help

4. The output is placed in a worksheet given the default name "Sheet1." You will want to widen column A of the output so that you can read the labels. Your output should be similar to the output shown below. The mean of the sample is 23.75, the median is 21.5, and the mode is 20.

Be careful when using the value of the mode reported in the Descriptive Statistics output. If there is a tie for the mode, Excel reports only the first value that occurs in the data set. Therefore, it is always a good idea to construct a frequency distribution table to go along with the Descriptive Statistics output.

	A	B
1	*Ages of students in Class*	
2		
3	Mean	23.75
4	Standard Error	2.193141
5	Median	21.5
6	Mode	20
7	Standard Deviation	9.808026
8	Sample Variance	96.19737
9	Kurtosis	19.0739
10	Skewness	4.324551
11	Range	45
12	Minimum	20
13	Maximum	65
14	Sum	475
15	Count	20

5. Construct a frequency distribution table to see if the data set is unimodal or multimodal. Go back to the worksheet containing the age data. To do this, click on the **ages** sheet tab near the bottom of the screen. Then select **PHStat → One-Way Tables & Charts.**

6. Complete the One-Way Tables & Charts dialog box as shown below. Click **OK**.

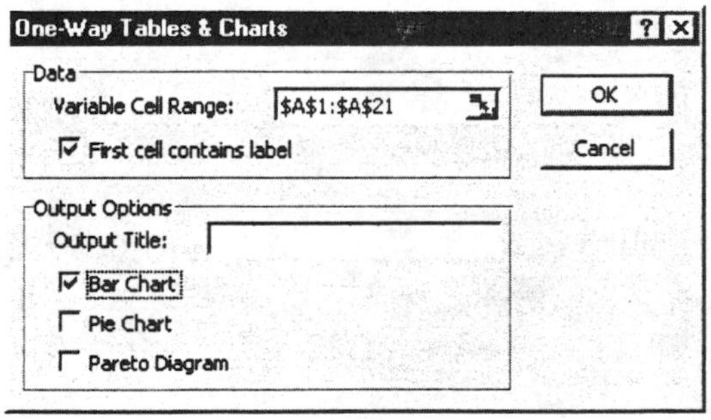

One-Way Tables & Charts

Data
Variable Cell Range: $A$1:$A$21
☑ First cell contains label

OK
Cancel

Output Options
Output Title:
☑ Bar Chart
☐ Pie Chart
☐ Pareto Diagram

The bar chart is displayed in a worksheet named "Bar Chart."

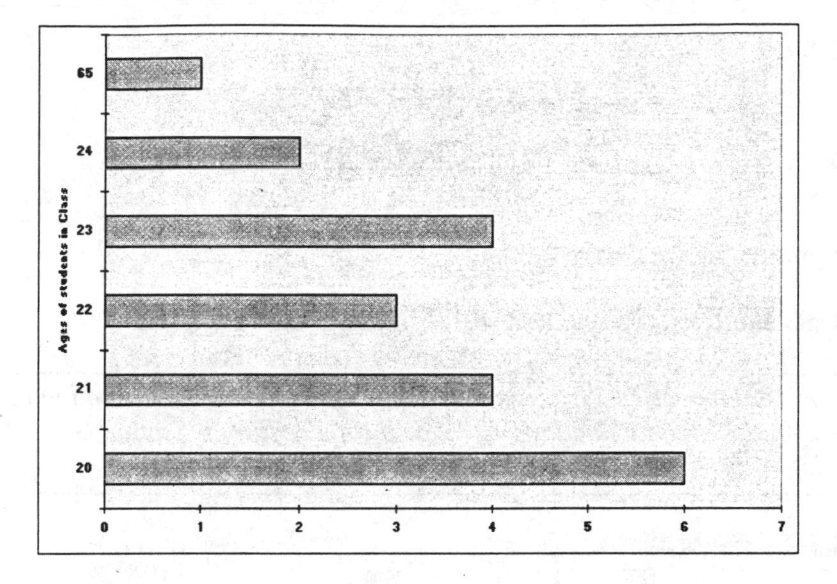

7. Click on the **One-Way Table** sheet tab at the bottom of the screen. This sheet contains a frequency distribution of the ages. You can see by looking at the bar chart and the frequency distribution that the age distribution is unimodal. The modal value of 20 has a frequency of six.

	A	B	C	D	E	F	G
1	One-Way Summary Table						
2							
3	Count of Ages of students in Class						
4	Ages of students in Class	Total					
5	20	6					
6	21	4					
7	22	3					
8	23	4					
9	24	2					
10	65	1					
11	Grand Total	20					

◀

Section 2.4

► Example 5 (pg. 72) — **Finding the Standard Deviation**

1. Open worksheet "atlanta" in the Chapter 2 folder.

2. Click **Tools → Data Analysis**. Select **Descriptive Statistics** and click **OK**.

If Data Analysis does not appear as a choice in the Tools menu, you will need to load the Microsoft Excel Analysis ToolPak add-in. Follow the procedure in Section GS 8.1 before continuing.

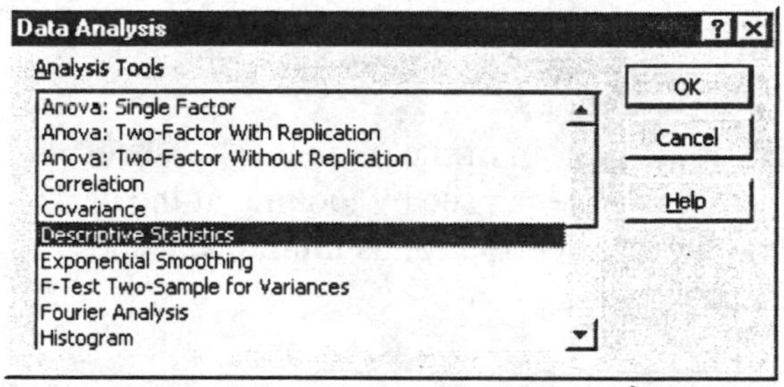

Data Analysis
Analysis Tools
Anova: Single Factor
Anova: Two-Factor With Replication
Anova: Two-Factor Without Replication
Correlation
Covariance
Descriptive Statistics
Exponential Smoothing
F-Test Two-Sample for Variances
Fourier Analysis
Histogram
OK
Cancel
Help

3. Complete the Descriptive Statistics dialog box as shown below. Click **OK**.

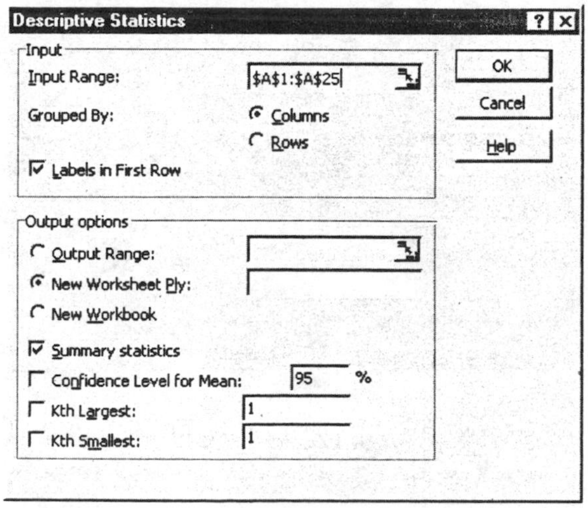

Descriptive Statistics
Input
Input Range: $A$1:$A$25
Grouped By: Columns / Rows
Labels in First Row
Output options
Output Range:
New Worksheet Ply:
New Workbook
Summary statistics
Confidence Level for Mean: 95 %
Kth Largest: 1
Kth Smallest: 1
OK
Cancel
Help

You will want to make Column A of the output wider so that you can read the labels. Your Descriptive Statistics output should be similar to the output shown below. The mean is 23.46875 and the standard deviation is 5.492244.

	A	B
1	*Atlanta rates*	
2		
3	Mean	23.46875
4	Standard Error	1.1211
5	Median	23.75
6	Mode	18
7	Standard Deviation	5.492244
8	Sample Variance	30.16474
9	Kurtosis	0.809987
10	Skewness	0.726594
11	Range	22.75
12	Minimum	14.25
13	Maximum	37
14	Sum	563.25
15	Count	24

Section 2.5

► Example 2 (pg. 86) Finding Quartiles

1. Open worksheet "tuition" in the Chapter 2 folder. Key in labels for the quartiles as shown below. Then click in cell **C2** to place the first quartile there.

	A	B	C	D	E	F	G
1	tuition		Quartile 1				
2	23						
3	25						
4	30		Quartile 2				
5	23						
6	20						
7	22		Quartile 3				

2. You will be using the QUARTILE function to obtain the first, second, and third quartiles for the tuition data. At the top of the screen, click **Insert → Function**.

3. Under Function category, select **Statistical** by clicking on it. Under Function name, select **QUARTILE**. Click **OK**.

Paste Function

Function category: Most Recently Used, All, Financial, Date & Time, Math & Trig, Statistical, Lookup & Reference, Database, Text, Logical, Information

Function name: POISSON, PROB, QUARTILE, RANK, RSQ, SKEW, SLOPE, SMALL, STANDARDIZE, STDEV, STDEVA

QUARTILE(array,quart)

Returns the quartile of a data set.

OK Cancel

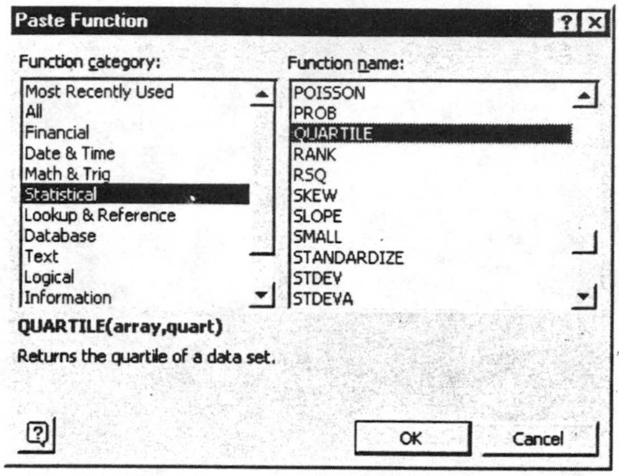

4. Complete the QUARTILE dialog box as shown below. Click **OK**.

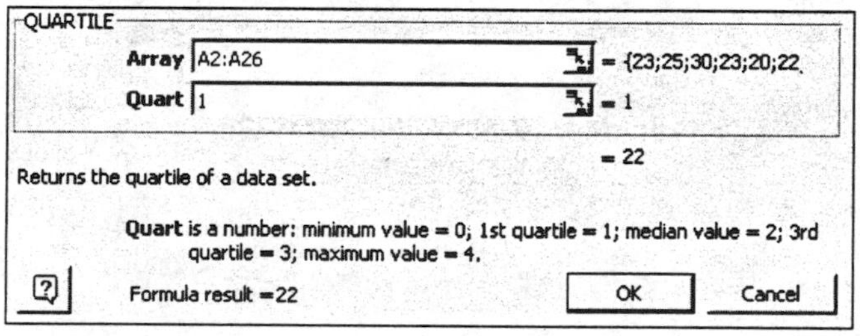

QUARTILE

Array A2:A26 = {23;25;30;23;20;22

Quart 1 = 1

= 22

Returns the quartile of a data set.

Quart is a number: minimum value = 0; 1st quartile = 1; median value = 2; 3rd quartile = 3; maximum value = 4.

Formula result =22

OK Cancel

5. Click in cell **C5** to place the second quartile there.

6. At the top of the screen, click **Insert → Function**.

7. Under Function category, select **Statistical** by clicking on it. Under Function name, select **QUARTILE**. Click **OK**.

8. Complete the QUARTILE dialog box as shown below. Click **OK**.

QUARTILE
Array A2:A26 = {23;25;30;23;20;22
Quart 2 = 2
= 23
Returns the quartile of a data set.
Quart is a number: minimum value = 0; 1st quartile = 1; median value = 2; 3rd quartile = 3; maximum value = 4.
Formula result =23 OK Cancel

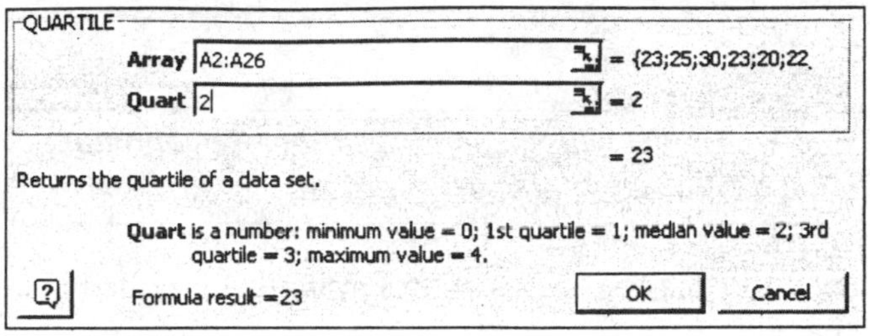

9. Click in cell **C8** to place the third quartile there.

10. At the top of the screen, click **Insert → Function**.

11. Under Function category, select **Statistical** by clicking on it. Under Function name, select **QUARTILE**. Click **OK**.

12. Complete the QUARTILE dialog box as shown below. Click **OK**.

QUARTILE
Array A2:A26 = {23;25;30;23;20;22
Quart 3 = 3
= 28
Returns the quartile of a data set.
Quart is a number: minimum value = 0; 1st quartile = 1; median value = 2; 3rd quartile = 3; maximum value = 4.
Formula result =28 OK Cancel

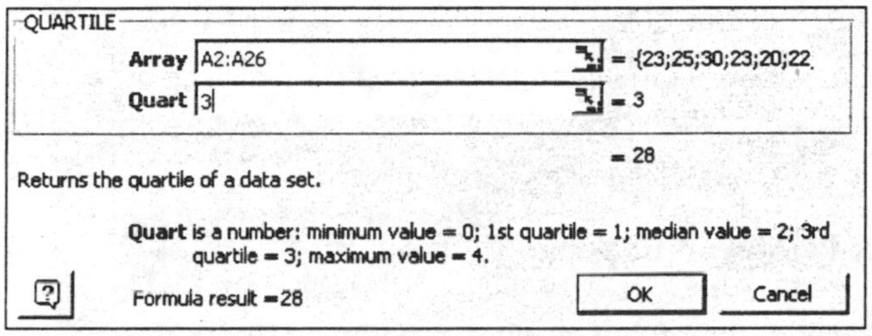

Your output should look similar to the output displayed below.

	A	B	C	D	E	F	G
1	tuition		Quartile 1				
2	23		22				
3	25						
4	30		Quartile 2				
5	23		23				
6	20						
7	22		Quartile 3				
8	21		28				

◀

► Example 4 (pg. 88) Drawing a Box-and-Whisker Plot

If the PHStat add-in has not been loaded, you will need to load it before continuing. Follow the instructions in Section GS 8.2.

1. Open a new Excel worksheet and enter the test scores of 15 employees enrolled in a CPR training course as shown below.

	A	B	C	D	E	F	G
1	Test score						
2	13						
3	9						
4	18						
5	15						
6	14						
7	21						
8	7						
9	10						
10	11						
11	20						
12	5						
13	18						
14	37						
15	16						
16	17						

2. At the top of the screen, select **PHStat → Box-and-Whisker Plot.**

3. Complete the Box-and-Whisker Plot dialog box as shown below. Click **OK**.

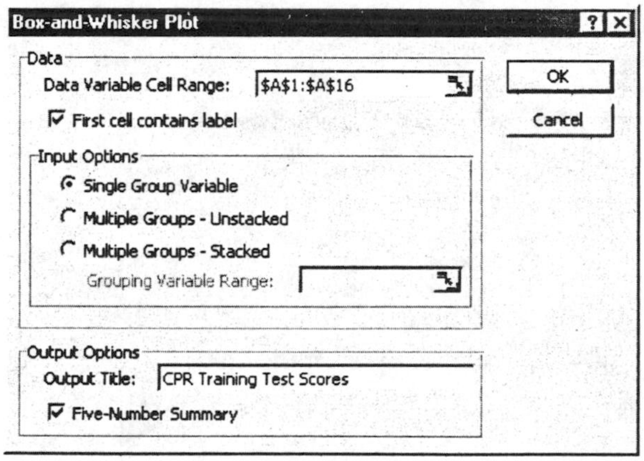

Box-and-Whisker Plot

Data
Data Variable Cell Range: $A$1:$A$16
☑ First cell contains label

Input Options
◉ Single Group Variable
○ Multiple Groups - Unstacked
○ Multiple Groups - Stacked
Grouping Variable Range:

Output Options
Output Title: CPR Training Test Scores
☑ Five-Number Summary

OK
Cancel

The Five-Number Summary is displayed in a worksheet named "FiveNumbers."

	A	B	C
1	CPR Training Test Scores		
2			
3	Five-number Summary		
4	Minimum	5	
5	First Quartile	10.5	
6	Median	15	
7	Third Quartile	18	
8	Maximum	37	

The box-and-whisker plot is displayed in a worksheet named "BoxWhiskerPlot."

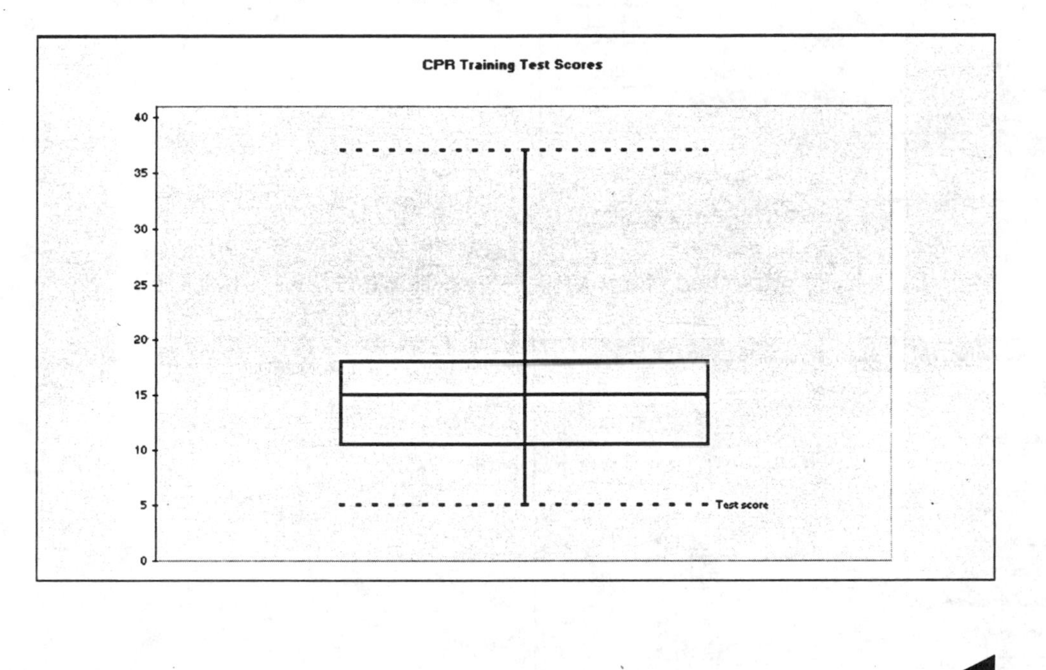

◀

Technology Lab

► Exercises 1 & 2 (pg. 93)

Finding the Sample Mean and the Sample Standard Deviation

1. Open worksheet "Tech2" in the Chapter 2 folder.

2. Click **Tools → Data Analysis**. Select **Descriptive Statistics** and click OK.

If Data Analysis does not appear as a choice in the Tools menu, you will need to load the Microsoft Excel Analysis ToolPak add-in. Follow the procedure in Section GS 8.1 before continuing.

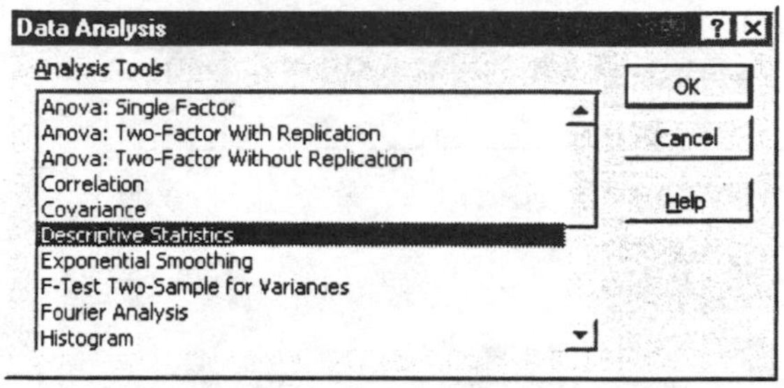

Data Analysis

Analysis Tools

Anova: Single Factor
Anova: Two-Factor With Replication
Anova: Two-Factor Without Replication
Correlation
Covariance
Descriptive Statistics
Exponential Smoothing
F-Test Two-Sample for Variances
Fourier Analysis
Histogram

OK
Cancel
Help

3. Complete the Descriptive Statistics dialog box as shown below. Click **OK**.

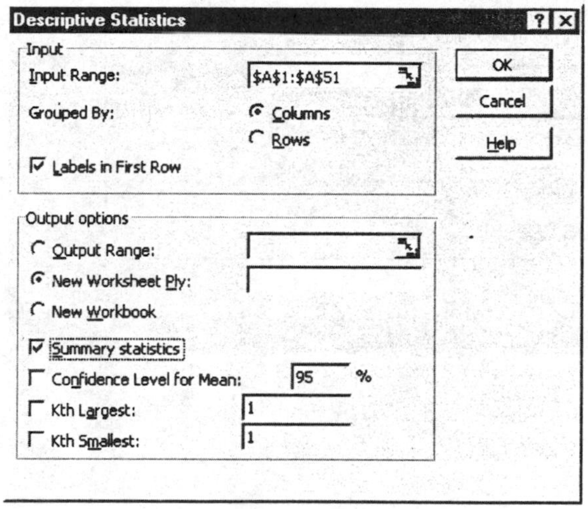

Descriptive Statistics

Input
Input Range: $A$1:$A$51
Grouped By: Columns / Rows
Labels in First Row

Output options
Output Range:
New Worksheet Ply:
New Workbook
Summary statistics
Confidence Level for Mean: 95 %
Kth Largest: 1
Kth Smallest: 1

OK
Cancel
Help

You will want to adjust the width of Column A of the output so that you can read the labels. Your Descriptive Statistics output should be similar to the output shown below. The sample mean is 2270.54 and the sample standard deviation is 653.1822.

	A	B
1	*Monthly Milk Production*	
2		
3	Mean	2270.54
4	Standard Error	92.37391
5	Median	2207
6	Mode	2207
7	Standard Deviation	653.1822
8	Sample Variance	426647
9	Kurtosis	0.567664
10	Skewness	0.549267
11	Range	3138
12	Minimum	1147
13	Maximum	4285
14	Sum	113527
15	Count	50

▸ Exercises 3 & 4 (pg. 93)

Constructing a Frequency Distribution and a Frequency Histogram

1. Open worksheet "Tech2" in the Chapter 2 folder.

2. Sort the data in ascending order. In the sorted data set you can see that the minimum production is 1147 pounds. Scroll down to find the maximum production. The maximum production, displayed in cell A51, is 4285 pounds.

For instructions on how to sort, refer to Sections GS 5.1 and GS 5.2.

	A	B	C	D	E	F	G
1	Monthly Milk Production						
2	1147						
3	1230						
4	1258						

3. Enter **Lower Limit** in cell D1 of the worksheet. The textbook tells you to use the minimum value as the lower limit of the first class. Enter **1147** in cell D2. You will calculate the remaining lower limits by adding the class width of 500 to the lower limit of each previous class. You will use a formula to do these computations in the Excel worksheet. Click in cell **D3** and key in **=D2+500** as shown below. Press [**Enter**].

	A	B	C	D	E	F	G
1	Monthly Milk Production			Lower Limit			
2	1147			1147			
3	1230			=D2+500			

4. Click on cell D3 (where 1647 now appears) and copy the contents of cell D3 to cells D4 through D9. Because the maximum milk production is 4285 pounds, you have calculated one more lower limit than is needed for the histogram. The value of 4647, however, will be used when calculating the upper limit of the last class.

	A	B	C	D	E	F	G
1	Monthly Milk Production			Lower Limit			
2	1147			1147			
3	1230			1647			
4	1258			2147			
5	1294			2647			
6	1319			3147			
7	1449			3647			
8	1619			4147			
9	1647			4647			

5. Enter **Upper Limit** in cell E1. The upper limit is equal to one less than the lower limit of the next higher class. To do these calculations, you will enter a formula in the Excel worksheet. Click in cell E2 and enter the formula **=D3-1** as shown in the worksheet below. Press [**Enter**].

	A	B	C	D	E	F	G
1	Monthly Milk Production			Lower Limit	Upper Limit		
2	1147			1147	=D3-1		

6. Copy the formula in E2 (where 1646 now appears) to cells E3 through E33. You will use these upper limits for the bins when you construct the histogram chart.

	A	B	C	D	E	F	G
1	Monthly Milk Production			Lower Limit	Upper Limit		
2	1147			1147	1646		
3	1230			1647	2146		
4	1258			2147	2646		
5	1294			2647	3146		
6	1319			3147	3646		
7	1449			3647	4146		
8	1619			4147	4646		
9	1647			4647			

7. At the top of the screen, click **Tools → Data Analysis**. Select **Histogram** and click **OK**.

If Data Analysis does not appear as a choice in the Tools menu, you will need to load the Microsoft Excel Analysis ToolPak add-in. Follow the procedure in Section GS 8.1 before continuing.

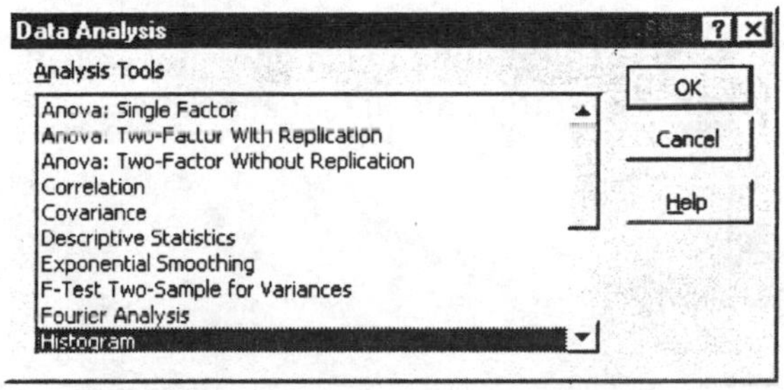

Data Analysis

Analysis Tools

Anova: Single Factor
Anova: Two-Factor With Replication
Anova: Two-Factor Without Replication
Correlation
Covariance
Descriptive Statistics
Exponential Smoothing
F-Test Two-Sample for Variances
Fourier Analysis
Histogram

OK
Cancel
Help

8. Complete the fields in the histogram dialog box as shown below. Click **OK**.

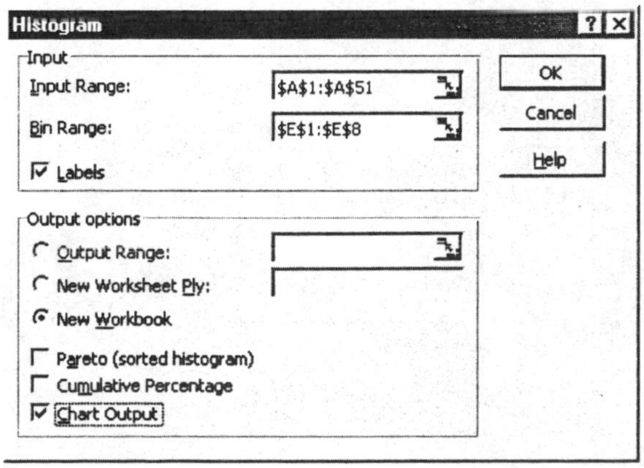

Histogram

Input
Input Range: $A$1:$A$51
Bin Range: $E$1:$E$8
☑ Labels

OK
Cancel
Help

Output options
○ Output Range:
○ New Worksheet Ply:
◉ New Workbook

☐ Pareto (sorted histogram)
☐ Cumulative Percentage
☑ Chart Output

9. You will now follow steps to modify the histogram so that it is presented in a more informative manner. Begin by making the chart taller so that it is easier to read. To do this, first click within the figure near a border. Black square handles appear. Click on the center handle at the bottom border of the figure and drag it down a few rows.

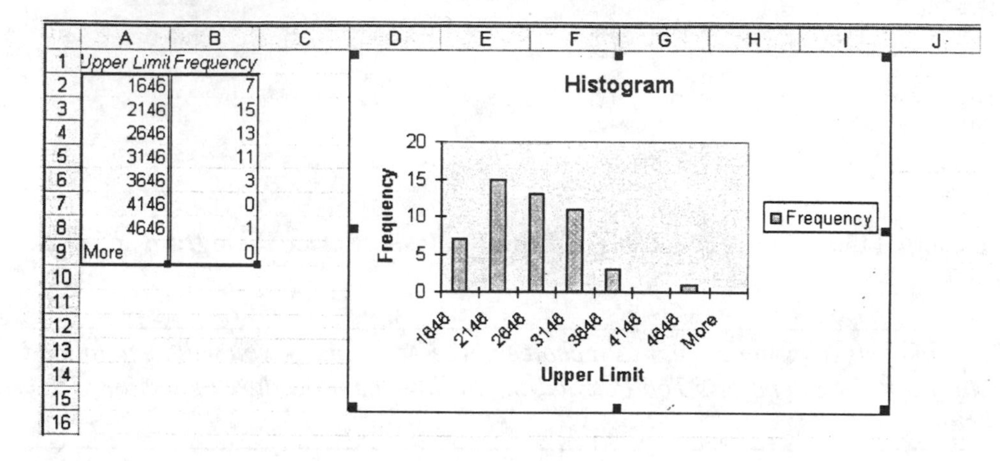

12. Remove the "More" category from the X axis. **Right click** in the gray plot area of the histogram and select **Source Data** from the shortcut menu that appears.

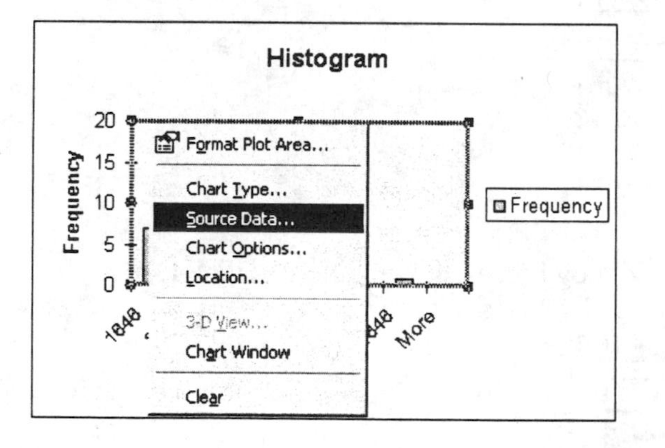

13. Click the **Series** tab at the top of the Source Data dialog box. "More" appears in cell A9 of the worksheet and its zero frequency appears in cell B9. To exclude the information in row 9 from the chart, edit the entry in the Values window and the entry in the Category (X) axis labels window. The entry in the Values window should read **=Sheet1!$B$2:$B$8**. The entry in the Category (X) axis labels window should read **=Sheet1!$A$2:$A$8**. Click **OK**.

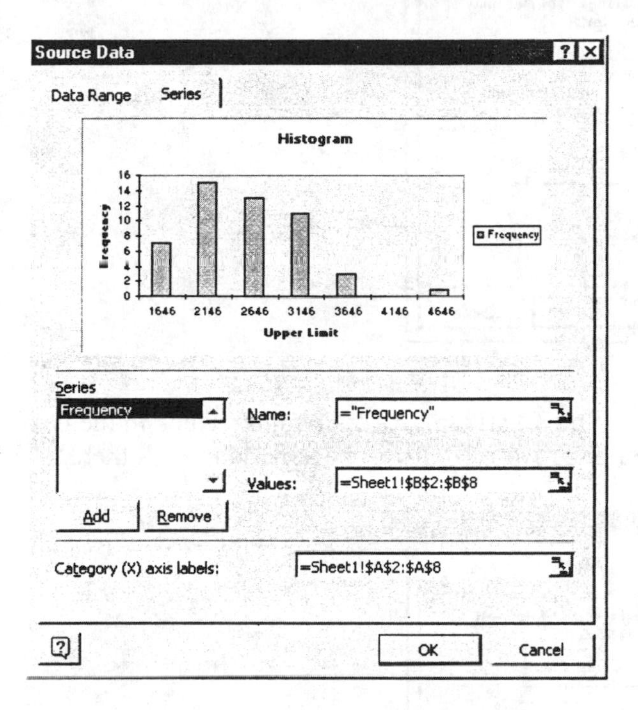

14. **Right click** in the gray plot area of the histogram and select **Chart Options** from the menu that appears.

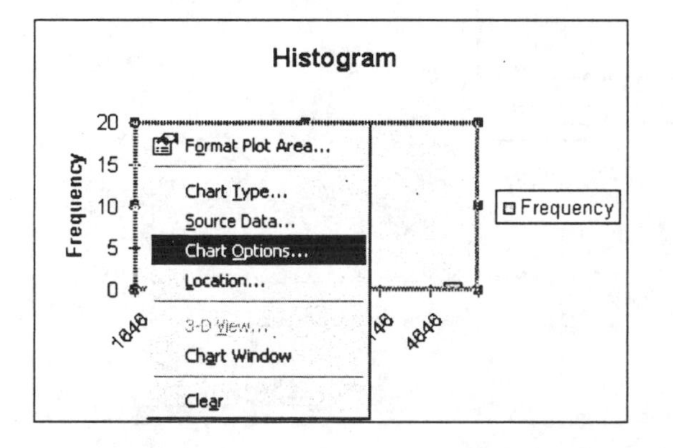

15. Click the **Titles** tab at the top of the Chart Options dialog box. In the Chart title field, replace "Histogram" with **Monthly Milk Production of 50 Holstein Dairy Cows**. In the Category (X) axis field, enter **Pounds**.

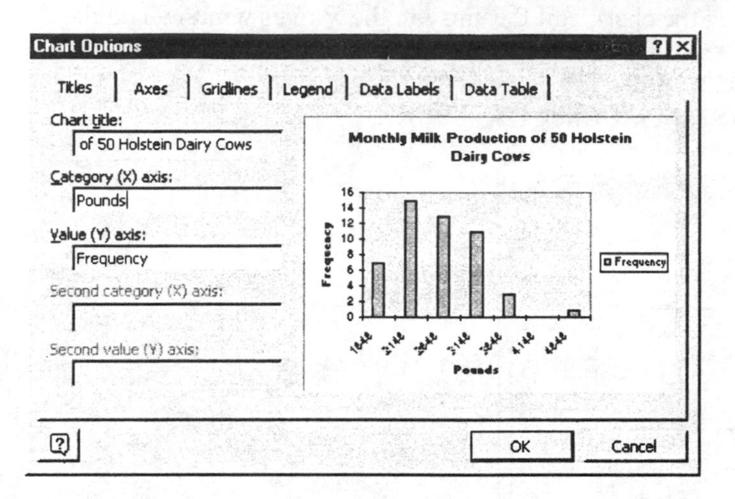

16. Click the **Gridlines** tab at the top of the Chart Options dialog box. Click in the box next to **Major gridlines** under Value (Y) axis so that a checkmark appears there.

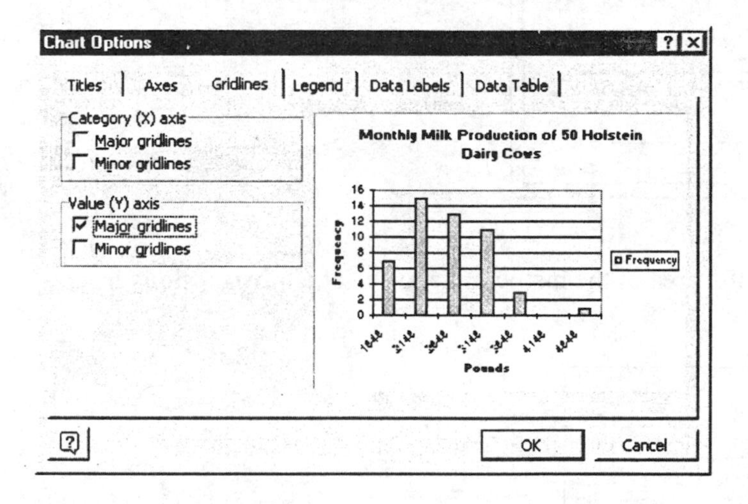

17. Click the **Legend** tab at the top of the Chart Options dialog box. Click in the box to the left of **Show Legend** to remove the checkmark. The removal of the checkmark will delete the frequency legend from the right side of the histogram chart. Click **OK**.

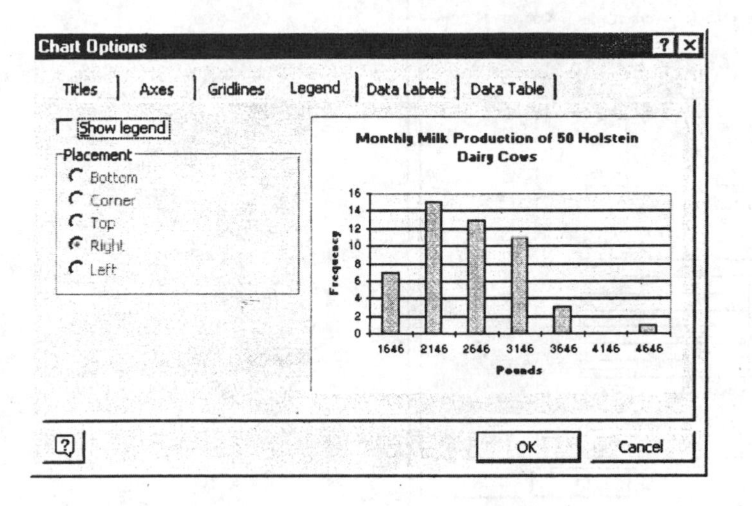

18. Remove the space between the vertical bars. **Right click** on one of the vertical bars. Select **Format Data Series** from the shortcut menu that appears.

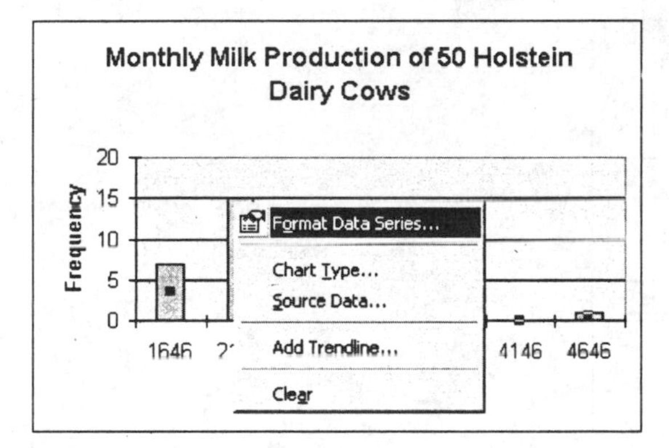

19. Click on the **Options** tab at the top of the Format Data Series dialog box. Change the value in the Gap width box to 0. Click **OK**.

Format Data Series

Patterns | Axis | Y Error Bars | Data Labels | Series Order | Options

Overlap: 0

Gap width: 0

Series lines

Vary colors by point

Monthly Milk Production of 50 Holstein Dairy Cows

Frequency

Pounds

OK Cancel

Your histogram should look similar to the one displayed below.

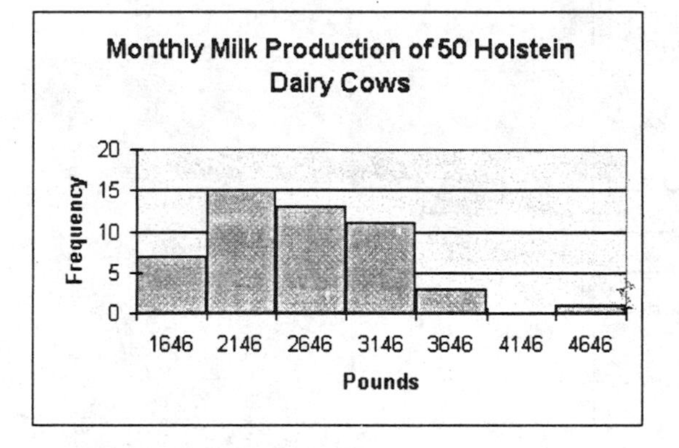

Probability

Section 3.2

► Exercise 25 (pg. 123) Simulating the "Birthday Problem"

You will be generating 24 random numbers between 1 and 365.

If the PHStat add-in has not been loaded, you will need to load it before continuing. Follow the instructions in Section GS 8.2.

1. Open a new Excel worksheet.

2. You will be entering the numbers 1 through 365 in the worksheet. To do this, you will fill column A with a number series. Begin by entering the numbers 1 and 2 in column A as shown below. You are starting a number series that will increase in increments of one.

	A	B	C	D	E	F	G
1	1						
2	2						

3. You will now complete the series all the way to 365. Begin by clicking in cell A1 and dragging down to cell A2 so that both cells are highlighted as shown below.

	A	B	C	D	E	F	G
1	1						
2	2						

4. Move the mouse pointer in cell A2 to the bottom right corner of that cell so that the white plus sign turns into a black plus sign. This black plus sign is called the "fill handle." While the fill handle is displayed, hold down on the left mouse button and

drag down column A until you reach cell A365. Release the mouse button. You should see the numbers from 1 to 365 in column A.

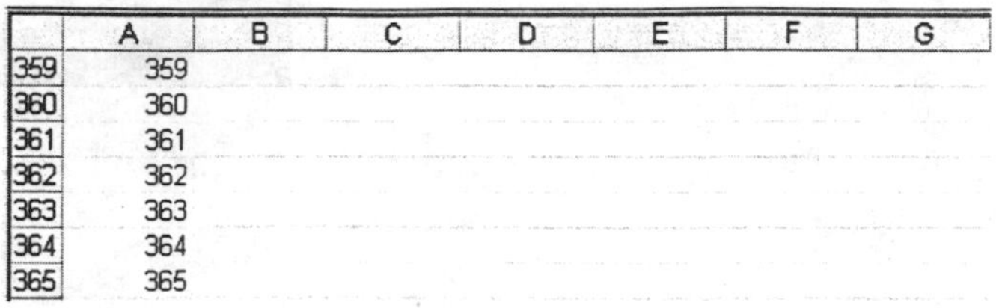

	A	B	C	D	E	F	G
359	359						
360	360						
361	361						
362	362						
363	363						
364	364						
365	365						

5. At the top of the screen, click **Tools** → **Data Analysis**. Select **Sampling** and click **OK**.

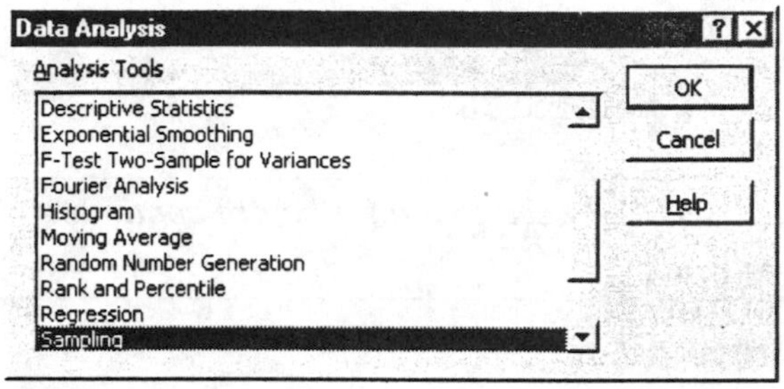

Data Analysis

Analysis Tools

Descriptive Statistics
Exponential Smoothing
F-Test Two-Sample for Variances
Fourier Analysis
Histogram
Moving Average
Random Number Generation
Rank and Percentile
Regression
Sampling

OK
Cancel
Help

6. Complete the Sampling dialog box as shown below. Click **OK**.

You will be generating 10 different samples of 24 numbers. The first set of 24 numbers will be placed in column B, the second in column C, etc.

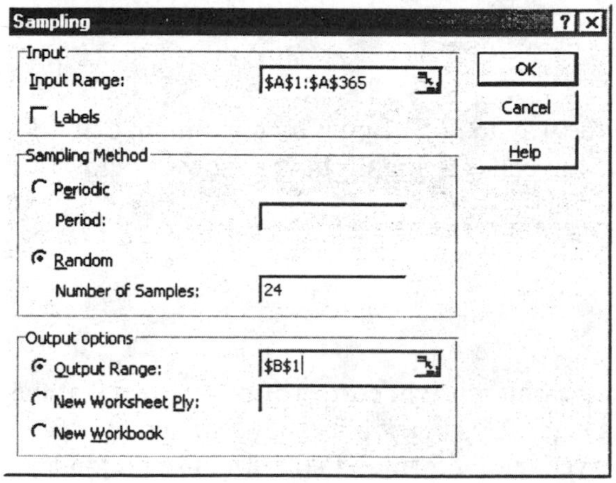

Sampling

Input
Input Range: $A$1:$A$365
Labels

Sampling Method
Periodic
Period:
Random
Number of Samples: 24

Output options
Output Range: $B$1
New Worksheet Ply:
New Workbook

OK
Cancel
Help

7. The 24 numbers, generated randomly with replacement, are displayed in column B. (Because the numbers were generated randomly, it is not likely that your output will be exactly the same.) Construct a frequency distribution to see if there are any repetitions. Select **PHStat → One-Way Tables and Charts**.

	A	B	C	D	E	F	G
1	1	308					
2	2	273					
3	3	120					
4	4	57					
5	5	350					
6	6	276					
7	7	259					
8	8	28					
9	9	51					
10	10	308					
11	11	7					
12	12	237					
13	13	60					
14	14	107					
15	15	171					
16	16	265					
17	17	353					
18	18	172					
19	19	211					
20	20	201					
21	21	89					
22	22	259					

8. Complete the One-Way Tables & Charts dialog box as shown below. Click **OK**.

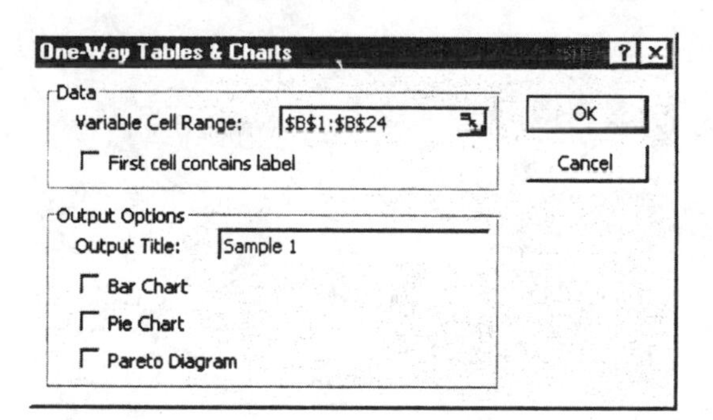

9. The frequency distribution table is displayed in a worksheet named "OneWayTable." In this example, you can see that the value of 259 has a frequency of 2.

	A	B
1	Sample 1	
2		
3	Count of Variable	
4	Variable	Total
5	7	1
6	28	1
7	51	1
8	57	1
9	60	1
10	89	1
11	107	1
12	120	1
13	171	1
14	172	1
15	194	1
16	201	1
17	211	1
18	237	1
19	259	2
20	265	1
21	273	1
22	276	1

Scrolling down, you can see that, in this example, the value of 308 also has a frequency of 2.

23	308	2
24	345	1
25	350	1
26	353	1
27	Grand Total	24

10. Generate the second sample. Go back to the sheet containing the numbers 1 through 365 by clicking on the **Sheet1** tab near the bottom of the screen. Then click **Tools → Data Analysis**. Select **Sampling** and click **OK**. Complete the Sampling dialog box as shown below. The second set of randomly generated numbers will be placed in column C. Click **OK**.

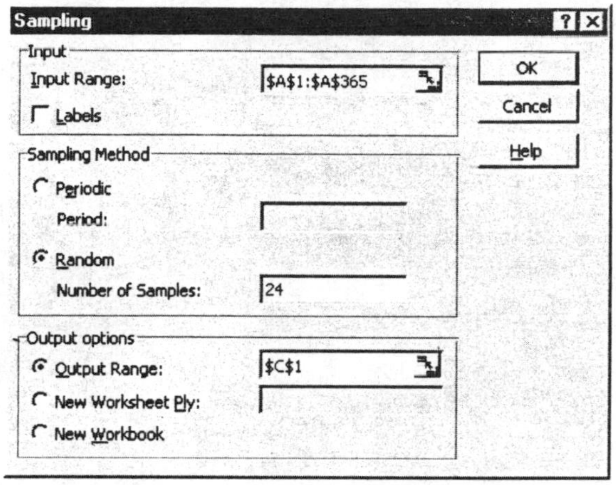

Sampling

Input
Input Range: $A$1:$A$365
Labels

Sampling Method
Periodic
Period:
Random
Number of Samples: 24

Output options
Output Range: $C$1
New Worksheet Ply:
New Workbook

OK
Cancel
Help

11. Construct a frequency distribution to see if there are any repetitions. Select **PHStat → One-Way Tables & Charts**. Complete the One-Way Tables & Charts dialog box as shown below. Click **OK**.

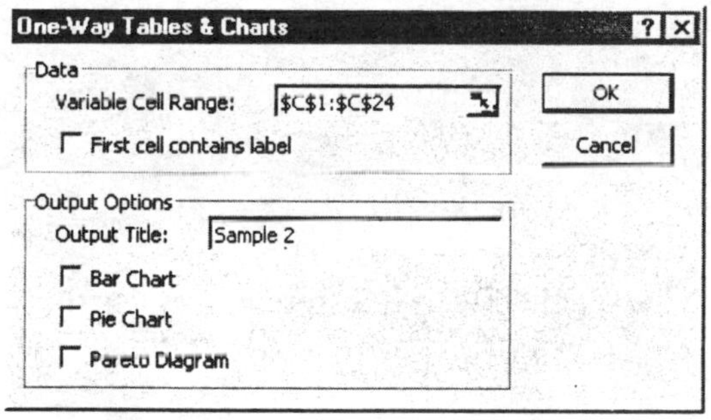

One-Way Tables & Charts

Data
Variable Cell Range: $C$1:$C$24
First cell contains label

Output Options
Output Title: Sample 2
Bar Chart
Pie Chart
Pareto Diagram

OK
Cancel

12. The frequency distribution table is displayed in a worksheet named "OneWayTable2." In this example, you can see that the value of 119 has a frequency of 2.

	A	B
1	Sample 2	
2		
3	Count of Variable	
4	Variable	Total
5	4	1
6	15	1
7	72	1
8	86	1
9	99	1
10	111	1
11	114	1
12	119	2
13	120	1
14	124	1
15	205	1
16	218	1
17	232	1
18	244	1
19	248	1
20	258	1
21	309	1
22	312	1

Scrolling down, you see no more repetitions in this example.

23	318	1
24	320	1
25	324	1
26	332	1
27	364	1
28	Grand Total	24

13. Repeat the appropriate steps until you have generated 10 different samples of 24 numbers.

◀

Section 3.4

► Example 3 (pg. 135) Finding the Number of Permutations of n Objects

1. Open a new Excel worksheet and click in cell **A1** to place the output there. You will be using the PERMUT function to calculate the number of permutations.

2. Click **Insert** → **Function**.

3. Under Function category, select **Statistical** by clicking on it. Under Function name, select **PERMUT**. Click **OK**.

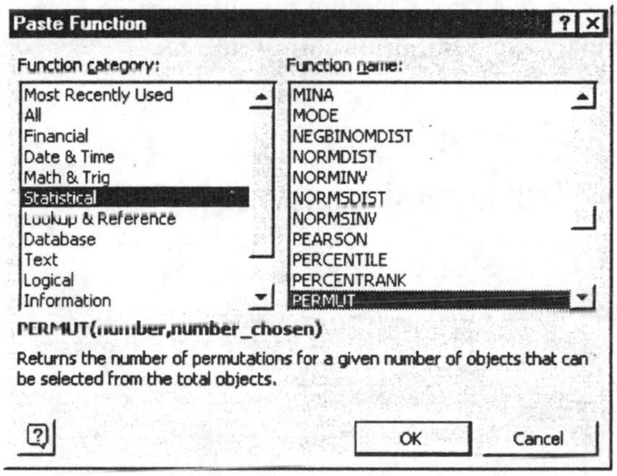

Paste Function

Function category: Most Recently Used, All, Financial, Date & Time, Math & Trig, Statistical, Lookup & Reference, Database, Text, Logical, Information

Function name: MINA, MODE, NEGBINOMDIST, NORMDIST, NORMINV, NORMSDIST, NORMSINV, PEARSON, PERCENTILE, PERCENTRANK, PERMUT

PERMUT(number,number_chosen)

Returns the number of permutations for a given number of objects that can be selected from the total objects.

OK Cancel

4. Complete the PERMUT dialog box as shown below. Click **OK**.

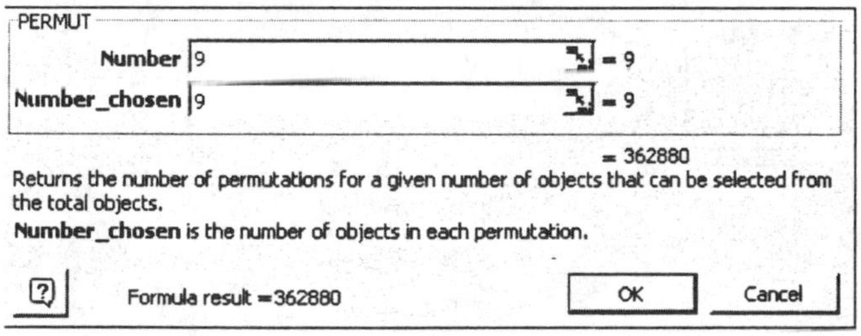

PERMUT

Number 9 = 9

Number_chosen 9 = 9

= 362880

Returns the number of permutations for a given number of objects that can be selected from the total objects.

Number_chosen is the number of objects in each permutation.

Formula result =362880 OK Cancel

The output is shown below. There are 362,880 different batting orders.

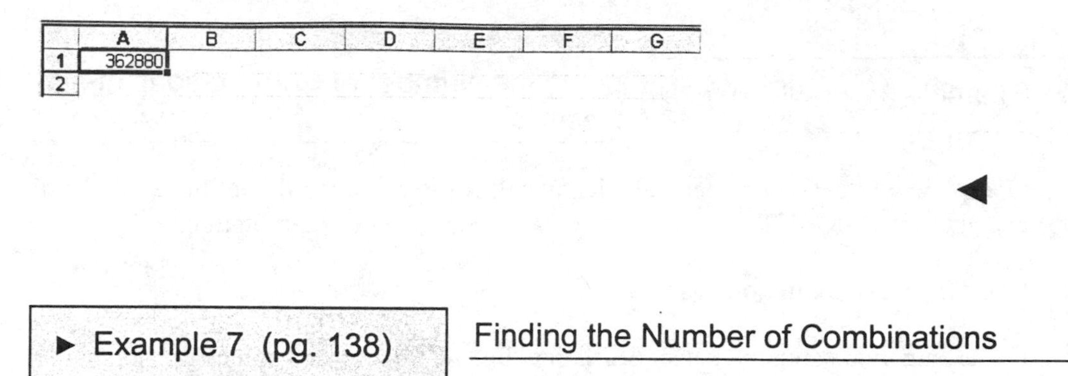

► Example 7 (pg. 138) Finding the Number of Combinations

1. Open a new Excel worksheet and click in cell **A1** to place the output there. You will be using the COMBIN function to calculate the number of combinations.

2. Click **Insert → Function.**

3. Under Function category, select **Math & Trig** by clicking on it. Under Function name, select **COMBIN**. Click **OK**.

Paste Function

Function category:
Most Recently Used
All
Financial
Date & Time
Math & Trig
Statistical
Lookup & Reference
Database
Text
Logical
Information

Function name:
ABS
ACOS
ACOSH
ASIN
ASINH
ATAN
ATAN2
ATANH
CEILING
COMBIN
COS

COMBIN(number,number_chosen)

Returns the number of combinations for a given number of items. See Help for the equation used.

OK Cancel

4. Complete the COMBIN dialog box as shown below. Click **OK**.

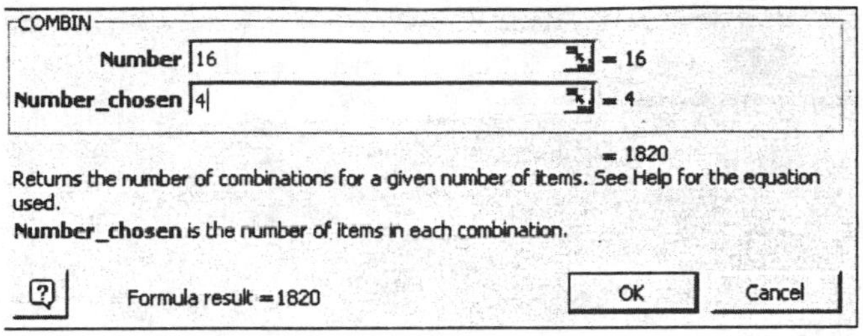

COMBIN
Number 16 = 16
Number_chosen 4 = 4
= 1820
Returns the number of combinations for a given number of items. See Help for the equation used.
Number_chosen is the number of items in each combination.
Formula result =1820
OK Cancel

The output is shown below. There are 1,820 different combinations.

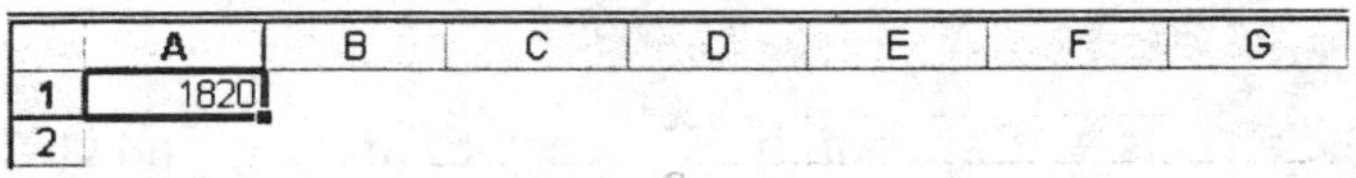

	A	B	C	D	E	F	G
1	1820						
2							

Technology Lab

► Exercise 3 (pg. 143)

Selecting Randomly a Number from 1 to 11

If the PHStat add-in has not been loaded, you will need to load it before continuing. Follow the instructions in Section GS 8.2.

You will be given the steps to follow to complete Exercise 3 and Exercise 3B.

1. You will first select randomly a number between 1 and 11. Open a new Excel worksheet. Enter the numbers 1 through 11 in cells A1 through A11 as shown below.

	A	B	C	D	E	F	G
1	1						
2	2						
3	3						
4	4						
5	5						
6	6						
7	7						
8	8						
9	9						
10	10						
11	11						

2. At the top of the screen, click **Tools → Data Analysis**. Select **Sampling**, and click **OK**.

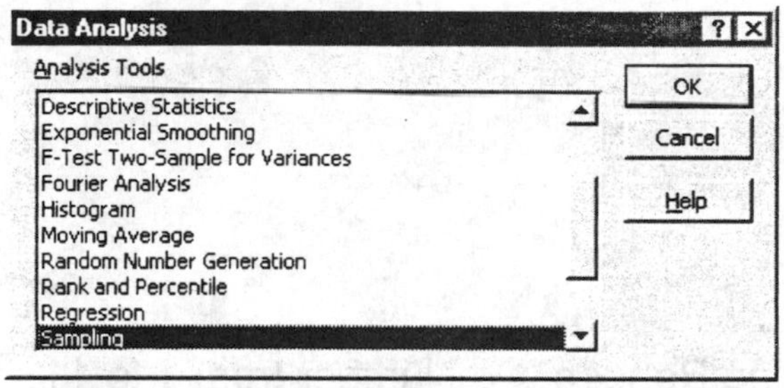

Data Analysis
Analysis Tools
Descriptive Statistics
Exponential Smoothing
F-Test Two-Sample for Variances
Fourier Analysis
Histogram
Moving Average
Random Number Generation
Rank and Percentile
Regression
Sampling
OK
Cancel
Help

3. Complete the Sampling dialog box as shown below. Click **OK**.

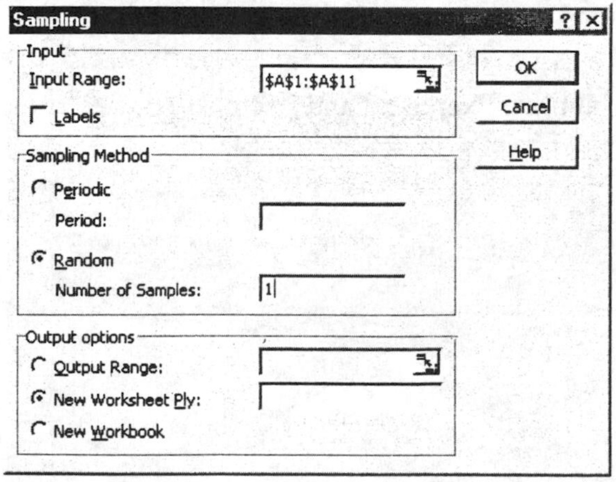

Sampling
Input
Input Range: $A$1:$A$11
Labels
Sampling Method
Periodic
Period:
Random
Number of Samples: 1
Output options
Output Range:
New Worksheet Ply:
New Workbook
OK
Cancel
Help

4. The output is displayed in a new sheet. The number 2 was selected. You will now select randomly 100 integers from 1 to 11. Click the **Sheet1** tab near the bottom of the screen to go back to the worksheet displaying the numbers from 1 to 11.

	A	B	C	D	E	F	G
1	2						
2							

5. Click **Tools** → **Data Analysis**. Select **Sampling** and click **OK**.

6. Complete the Sampling dialog box as shown below. Click **OK**.

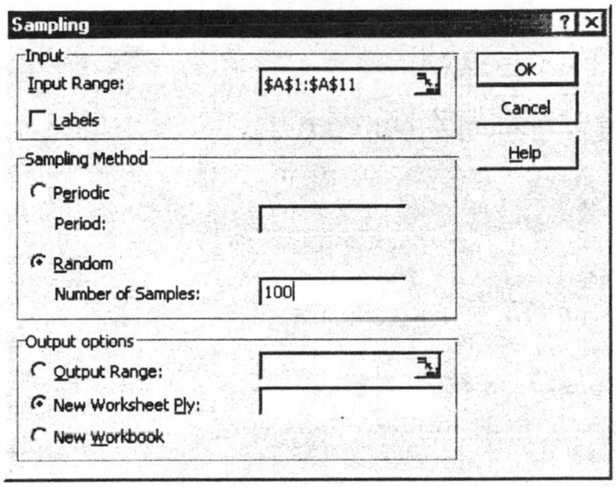

Sampling

Input
Input Range: $A$1:$A$11
Labels

Sampling Method
Periodic
Period:
Random
Number of Samples: 100

Output options
Output Range:
New Worksheet Ply:
New Workbook

OK
Cancel
Help

7. The output is displayed in a new sheet. You will need to scroll down to see the entire listing of the 100 randomly generated integers. You will now tally the results. Select **PHStat** → **One-Way Tables & Charts**.

	A	B	C	D	E	F	G
1	4						
2	1						
3	7						
4	2						
5	8						
6	8						
7	2						

8. Complete the One-Way Tables & Charts dialog box as shown below. Click **OK**.

One-Way Tables & Charts

Data
Variable Cell Range: $A$1:$A$100
☐ First cell contains label

OK
Cancel

Output Options
Output Title:
☐ Bar Chart
☐ Pie Chart
☐ Pareto Diagram

The output for Exercise 3B is displayed below. Your output should have the same format. Because the numbers were generated randomly, however, it is not likely that your numbers will be exactly the same.

	A	B	C
1	One-Way Summary Table		
2			
3	Count of Variable		
4	Variable	Total	
5	1	11	
6	2	14	
7	3	8	
8	4	8	
9	5	6	
10	6	6	
11	7	10	
12	8	13	
13	9	8	
14	10	6	
15	11	10	
16	Grand Total	100	

◄

► Exercise 5 (pg. 143) Composing Mozart Variations with Dice

If the PHStat add-in has not been loaded, you will need to load it before continuing. Follow the instructions in Section GS 8.2.

You will first be given the steps to follow to complete Exercise 5 and then the steps to follow to complete Exercise 5B. In Exercise 5, you will select randomly two integers from 1, 2, 3, 4, 5, and 6. This simulates tossing two six-sided dice one time. You will then sum the two integers and subtract 1 from the sum. This is Mozart's procedure (described at the top of page 143) for selecting a musical phrase from 11 different choices.

1. To select randomly two integers between 1 and 6, begin by opening a new Excel worksheet and entering the numbers 1 through 6 in column A as shown below.

	A	B	C	D	E	F	G
1	1						
2	2						
3	3						
4	4						
5	5						
6	6						

2. At the top of the screen, click **Tools → Data Analysis**. Select **Sampling** and click **OK**.

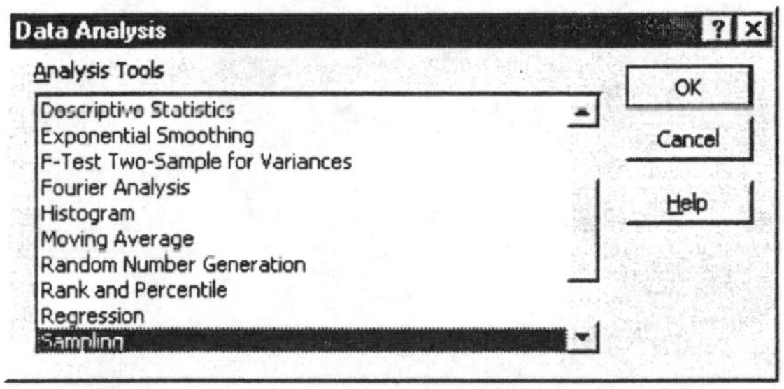

Data Analysis
Analysis Tools
Descriptive Statistics
Exponential Smoothing
F-Test Two-Sample for Variances
Fourier Analysis
Histogram
Moving Average
Random Number Generation
Rank and Percentile
Regression
Sampling
OK
Cancel
Help

3. Complete the Sampling dialog box as shown below. Click **OK**.

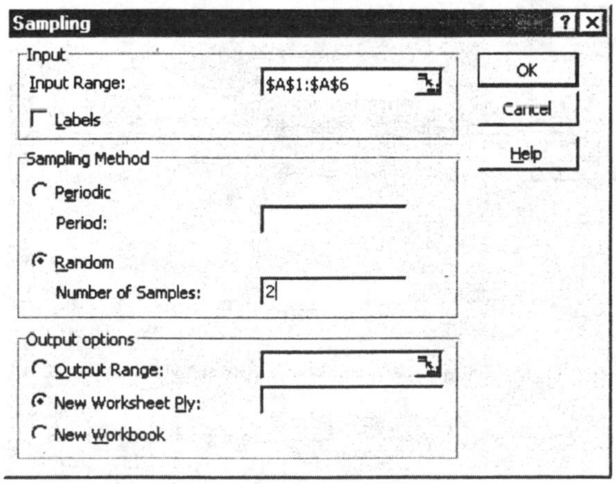

Sampling
Input
Input Range: $A$1:$A$6
Labels
Sampling Method
Periodic
Period:
Random
Number of Samples: 2
Output options
Output Range:
New Worksheet Ply:
New Workbook
OK
Cancel
Help

4. The output is displayed in a new worksheet. The numbers 5 and 3 were selected. Mozart's instructions are to sum the numbers and subtract 1. The result is 7. Therefore, musical phrase number 7 would be selected for the first bar of Mozart's minuet.

	A	B	C	D	E	F	G
1	5						
2	3						

5. Exercise 5B asks you to select 100 integers between 1 and 11. This simulates, 100 times, Mozart's procedure of tossing two die, finding the sum, and subtracting 1. You are also asked to tally the results. Begin by entering the numbers 1 through 11 in a new worksheet as shown below.

	A	B	C	D	E	F	G
1	1						
2	2						
3	3						
4	4						
5	5						
6	6						
7	7						
8	8						
9	9						
10	10						
11	11						

6. Click **Tools** → **Data Analysis**. Select **Sampling** and click **OK**.

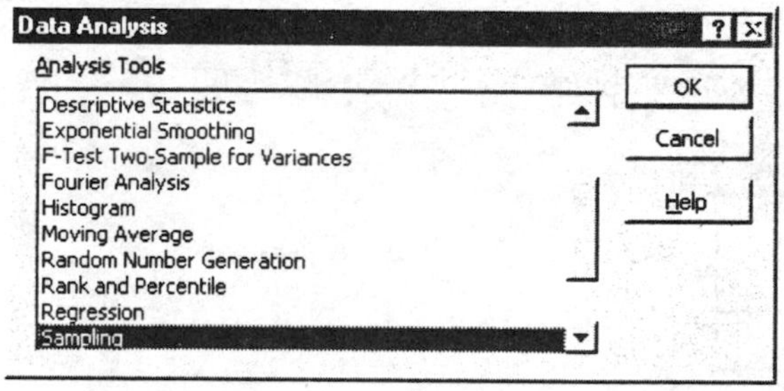

Data Analysis

Analysis Tools

Descriptive Statistics
Exponential Smoothing
F-Test Two-Sample for Variances
Fourier Analysis
Histogram
Moving Average
Random Number Generation
Rank and Percentile
Regression
Sampling

OK

Cancel

Help

7. Complete the Sampling dialog box as shown below. Click **OK**.

Sampling ? X

Input
Input Range: $A$1:$A$11
Labels

Sampling Method
Periodic
Period:
Random
Number of Samples: 100

Output options
Output Range:
New Worksheet Ply:
New Workbook

OK
Cancel
Help

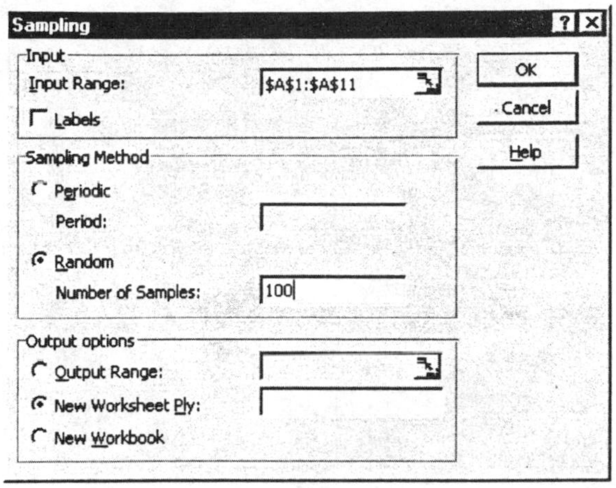

8. The output is displayed in a new sheet. You need to scroll down to see the entire listing of the 100 randomly generated integers. You will now tally the results. Select **PHStat → One-Way Tables & Charts.**

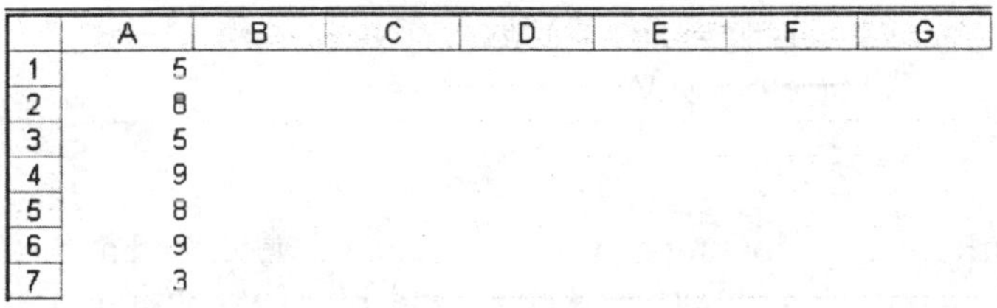

	A	B	C	D	E	F	G
1	5						
2	8						
3	5						
4	9						
5	8						
6	9						
7	3						

9. Complete the One-Way Tables and Charts dialog box as shown below. Click **OK**.

One-Way Tables & Charts ? X

Data
Variable Cell Range: $A$1:$A$100
First cell contains label

Output Options
Output Title:
Bar Chart
Pie Chart
Pareto Diagram

OK
Cancel

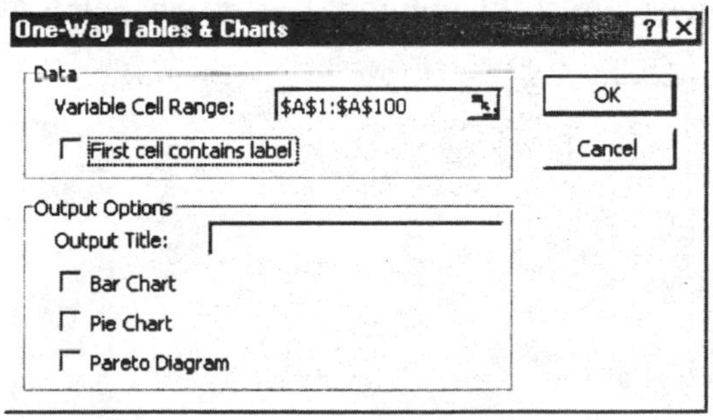

10. The output from Exercise 5B is displayed below. Because the numbers were generated randomly, it unlikely that your numbers will be exactly the same.

	A	B	C
1	One-Way Summary Table		
2			
3	Count of Variable		
4	Variable	Total	
5	1	10	
6	2	8	
7	3	11	
8	4	6	
9	5	15	
10	6	9	
11	7	4	
12	8	12	
13	9	10	
14	10	10	
15	11	5	
16	Grand Total	100	

◄

► Exercise 7 (pg. 143) Choosing a Minuet

You will be constructing a 16-bar minuet following Mozart's procedure described at the top of page 143. You first will generate 16 numbers between 1 and 11. Two options are available for the eighth bar and the sixteenth bar. Mozart said to choose Option 1 if the dice total is odd and to choose Option 2 if the dice total is even.

1. Open a new worksheet and enter the numbers 1 to 11 in column A as shown below.

	A	B	C	D	E	F	G
1	1						
2	2						
3	3						
4	4						
5	5						
6	6						
7	7						
8	8						
9	9						
10	10						
11	11						

2. At the top of the screen, click **Tools** → **Data Analysis**. Select **Sampling** and click **OK**.

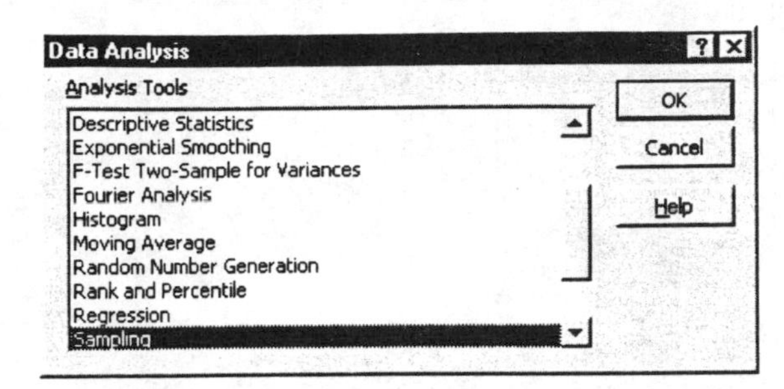

3. Complete the Sampling dialog box as shown below. Click **OK**.

Sampling
Input
Input Range: $A$1:$A$11
Labels
Sampling Method
Periodic
Period:
Random
Number of Samples: 16
Output options
Output Range:
New Worksheet Ply:
New Workbook
OK
Cancel
Help

The output is displayed in a new worksheet. The number for the eighth bar is 1. Because 1 is odd, you will select Option 1. The number for the sixteenth bar is 6, which is even. Option 2 will be selected for the sixteenth bar.

	A	B	C	D	E	F	G
1	11						
2	10						
3	9						
4	7						
5	2						
6	2						
7	8						
8	1	Option 1					
9	11						
10	6						
11	4						
12	3						
13	11						
14	6						
15	6						
16	6	Option 2					

You are now ready to go to the Internet site given in your textbook to play this minuet.

◀

Discrete Probability Distributions

Section 4.2

► Example 5 (pg. 169) — Finding a Binomial Probability

1. Open a new Excel worksheet and click in cell **A1** to place the output there.

2. Click **Insert** → **Function**.

3. Under Function category, select **Statistical**. Under Function name, select **BINOMDIST**. Click **OK**.

Paste Function

Function category:	Function name:
Most Recently Used	AVEDEV
All	AVERAGE
Financial	AVERAGEA
Date & Time	BETADIST
Math & Trig	BETAINV
Statistical	BINOMDIST
Lookup & Reference	CHIDIST
Database	CHIINV
Text	CHITEST
Logical	CONFIDENCE
Information	CORREL

BINOMDIST(number_s,trials,probability_s,cumulative)
Returns the individual term binomial distribution probability.

OK Cancel

4. Complete the BINOMDIST dialog box as shown below. Click **OK**.

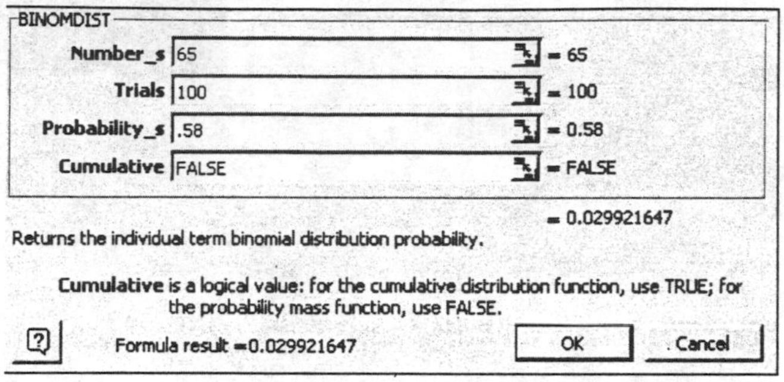

BINOMDIST

Number_s	65	= 65
Trials	100	= 100
Probability_s	.58	= 0.58
Cumulative	FALSE	= FALSE

= 0.029921647

Returns the individual term binomial distribution probability.

Cumulative is a logical value: for the cumulative distribution function, use TRUE; for the probability mass function, use FALSE.

? Formula result =0.029921647 OK Cancel

The BINOMDIST function returns a result of 0.029922.

	A	B	C	D	E	F	G
1	0.029922						
2							

◀

▶ Example 7 (pg. 171)

Constructing and Graphing a Binomial Distribution

1. Open a new, blank Excel worksheet and enter the information shown below. You will be using the BINOMDIST function to calculate binomial probabilities for 0, 1, 2, 3, 4, 5, and 6 households.

	A	B	C	D	E	F	G
1	Households	Relative frequency					
2	0						
3	1						
4	2						
5	3						
6	4						
7	5						
8	6						

2. Click in cell **B2** below "Relative frequency." Click **Insert→ Function**.

3. Under Function category, select **Statistical**. Under Function name, select **BINOMDIST**. Click **OK**.

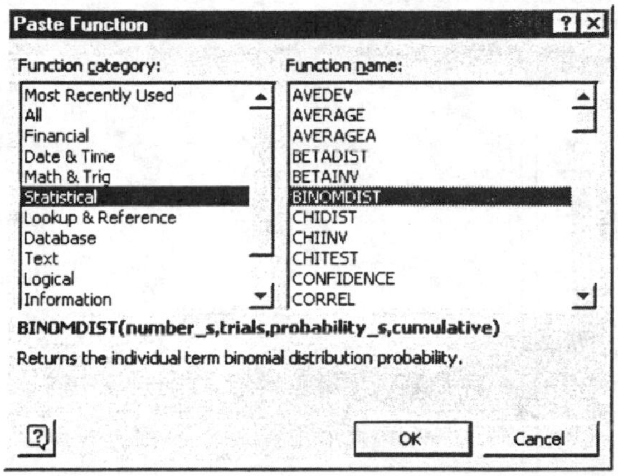

Paste Function

Function category:	Function name:
Most Recently Used	AVEDEV
All	AVERAGE
Financial	AVERAGEA
Date & Time	BETADIST
Math & Trig	BETAINV
Statistical	BINOMDIST
Lookup & Reference	CHIDIST
Database	CHIINV
Text	CHITEST
Logical	CONFIDENCE
Information	CORREL

BINOMDIST(number_s,trials,probability_s,cumulative)

Returns the individual term binomial distribution probability.

OK Cancel

4. Complete the BINOMDIST dialog box as shown below. Click **OK**.

You are entering a relative cell address without dollar signs (i.e., A2) in the Number_s field because you will be copying the contents of cell B2 to cells B3 through B8. You want the column A cell address to change from A2 to A3, A4, A5, ..., A8 when the formula is copied from cell B2 to cells B3 through B8.

BINOMDIST

Number_s	A2	= 0
Trials	6	= 6
Probability_s	.65	= 0.65
Cumulative	FALSE	= FALSE

= 0.001838266

Returns the individual term binomial distribution probability.

Cumulative is a logical value: for the cumulative distribution function, use TRUE; for the probability mass function, use FALSE.

Formula result =0.001838266 OK Cancel

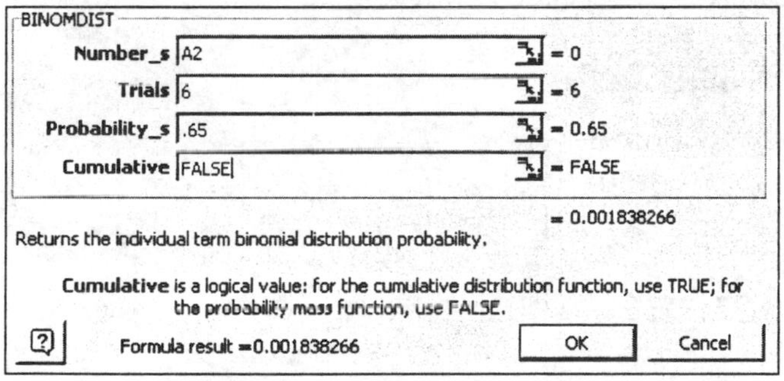

5. Copy the contents of cell B2 to cells B3 through B8.

	A	B	C	D	E	F	G
1	Households	Relative frequency					
2	0	0.001838					
3	1	0.020484					
4	2	0.095102					
5	3	0.235491					
6	4	0.328005					
7	5	0.243661					
8	6	0.075419					

6. Click in any cell in the data table. Then click **Insert → Chart**.

7. Under Chart type, in the Chart Type dialog box, select **Column**. Under Chart sub-type, select the first one in the top row by clicking on it. Click **Next**.

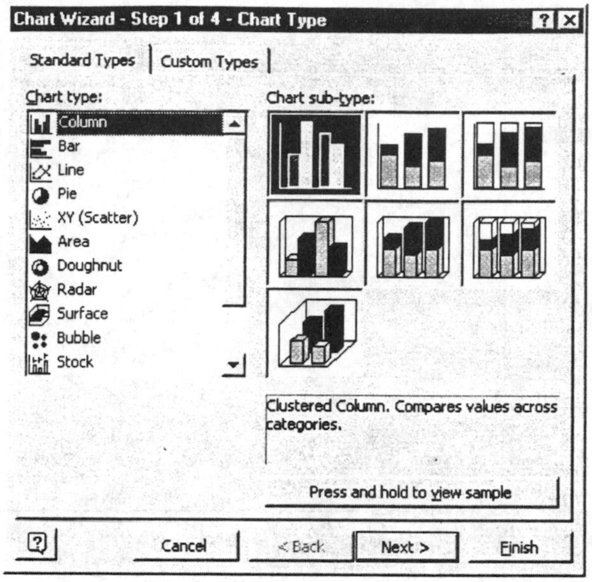

8. Check the accuracy of the data range in the Chart Source Data dialog box. It should read **=Sheet1!$A$1:$B$8**. Make any necessary corrections. Then click **Next**.

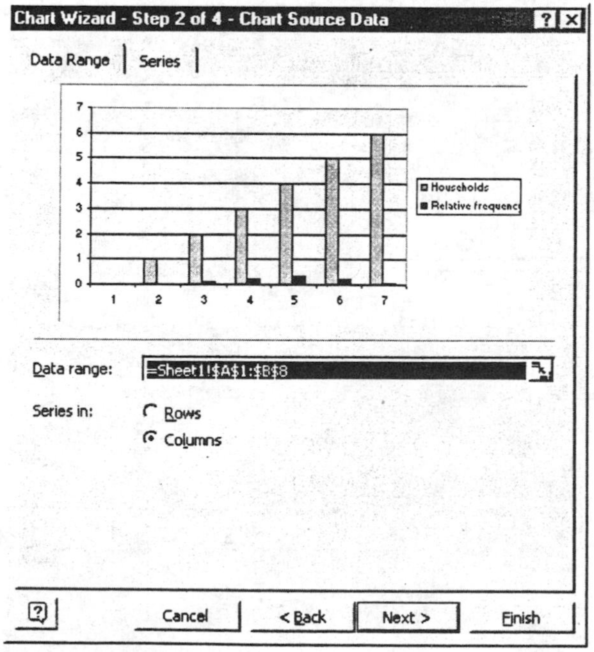

9. Click the **Titles** tab at the top of the Chart Options dialog box. In the Chart title window, enter **Subscribing to Cable TV**. In the Category (X) axis window, enter **Households**. In the Value (Y) axis window, enter **Relative Frequency**.

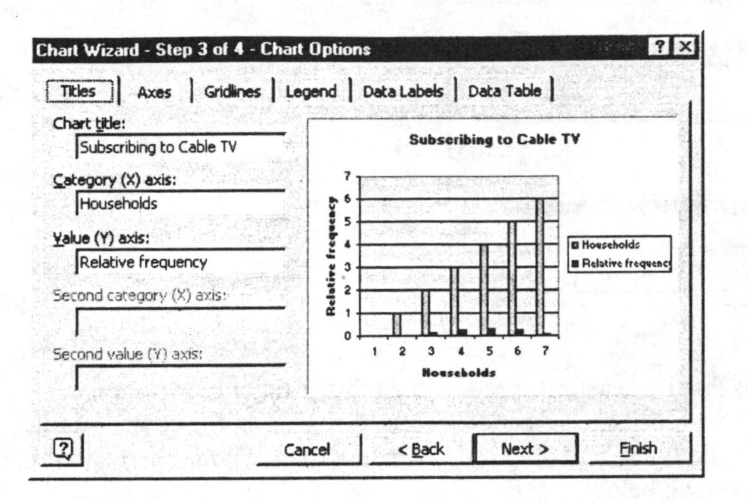

10. Click the **Legend** tab at the top of the Chart Options dialog box. Click in the box to the left of **Show legend** so that no checkmark appears there. This will remove the Households and Relative frequency legends on the right side of the chart. Click **Next**.

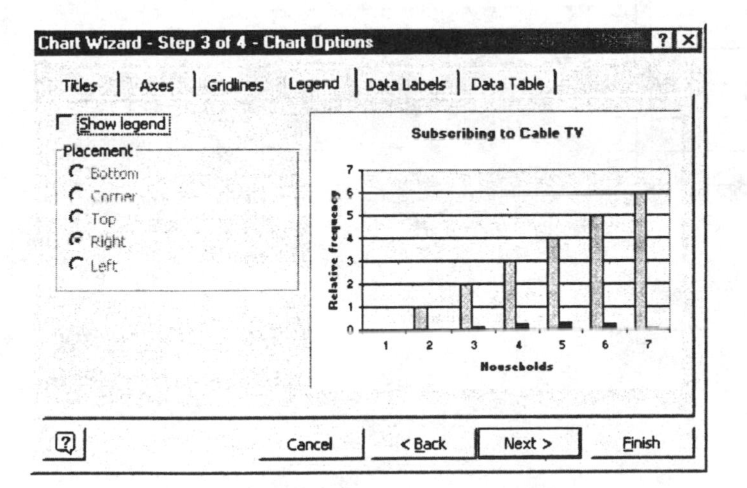

11. The Chart Location dialog box presents two options for placement of the chart. For this example, select **As object in**. Click **Finish**.

Chart Wizard - Step 4 of 4 - Chart Location

Place chart:

As new sheet: Chart1

As object in: Sheet1

Cancel | < Back | Next > | Finish

12. Make the chart taller so that it is easier to read. To do this, first click within the figure near a border so that black square handles appear. Click on the center handle on the bottom border of the figure and drag it down a few rows. Your chart should look similar to the one shown below.

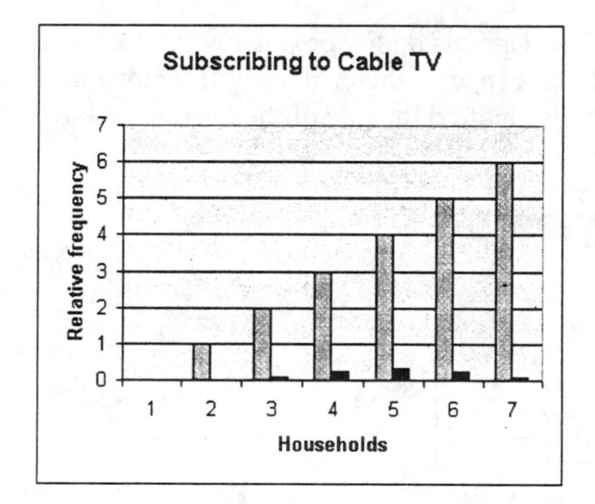

13. Correct the number scales displayed on the Y-axis and the X-axis. **Right click** on one of the vertical bars. Select **Source Data** from the shortcut menu that appears.

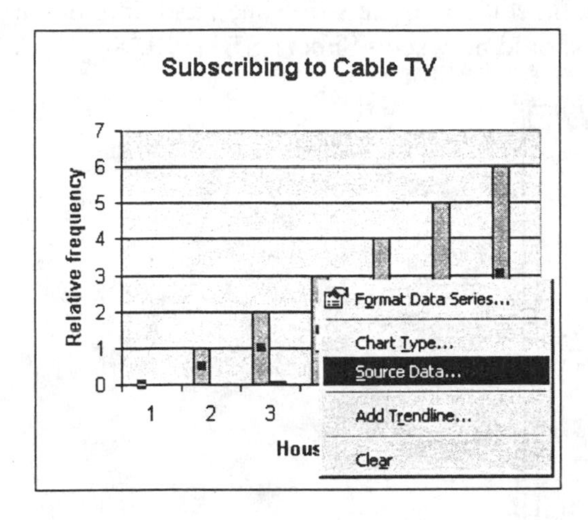

14. Click the **Series** tab at the top of the Source Data dialog box. In the Series window, select **Relative Frequency** by clicking on it. Then click the **Remove** button below the Series window.

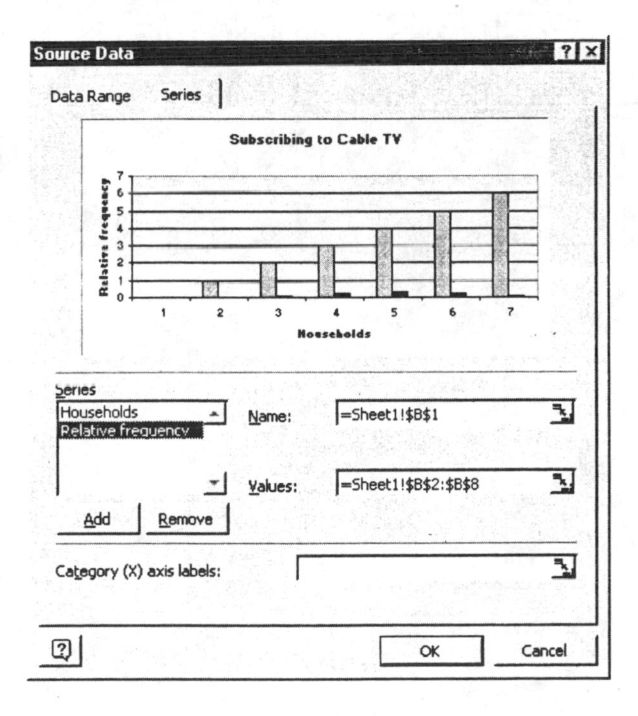

15. The Y-axis values are displayed in column B, so the entry in the **Values** window should be **=Sheet1!$B$2:$B$8**. First delete the information in the Values window. Next, click in the Values window, click on cell B2 of the worksheet, and drag down to cell B8. In the Values window, you should now see **=Sheet1!$B$2:$B$8**.

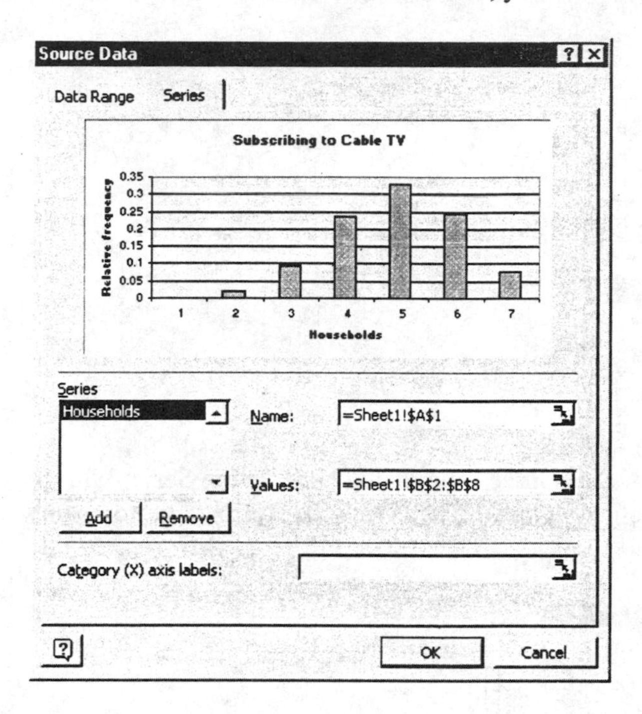

16. Column A contains the numbers you want displayed on the X-axis. Click in the Category (X) axis labels window. Then click on cell A2 of the worksheet and drag down to cell A8. The entry in the Category (X) axis labels field should now read **=Sheet1!$A$2:$A$8**. Click **OK**.

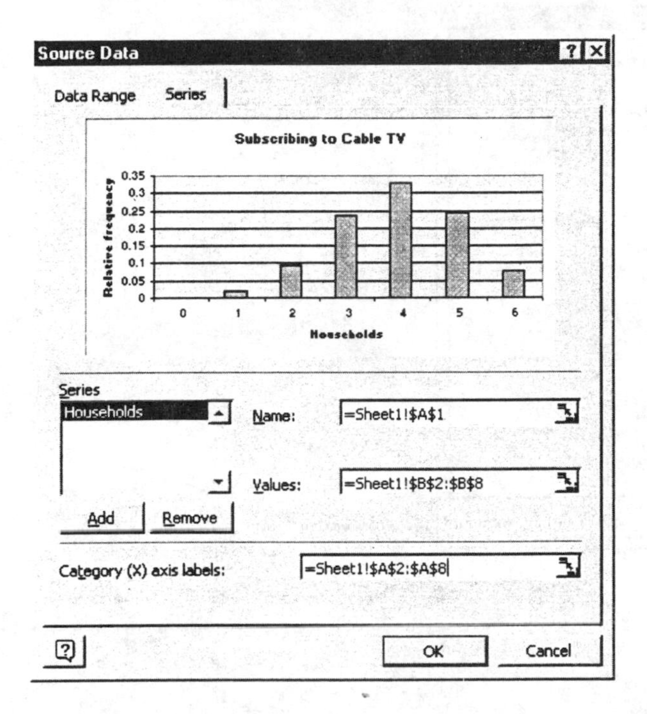

17. Remove the space between the vertical bars. **Right click** on one of the vertical bars. Select **Format Data Series** from the shortcut menu that appears.

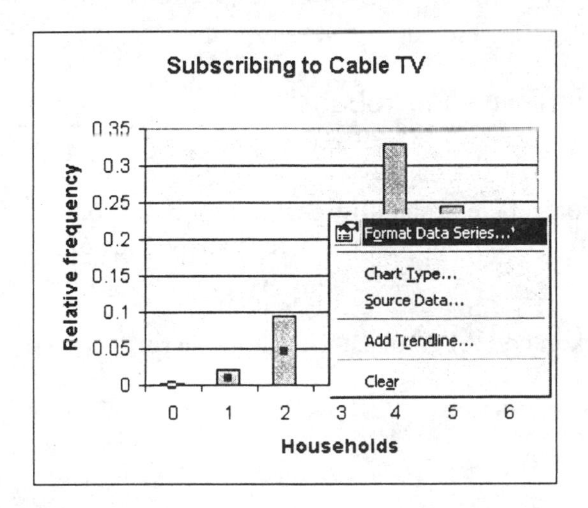

18. Click the **Options** tab at the top of the Format Data Series dialog box. Change the value in the Gap width box to 0. Click **OK**. Your relative frequency histogram should look similar to the one shown below.

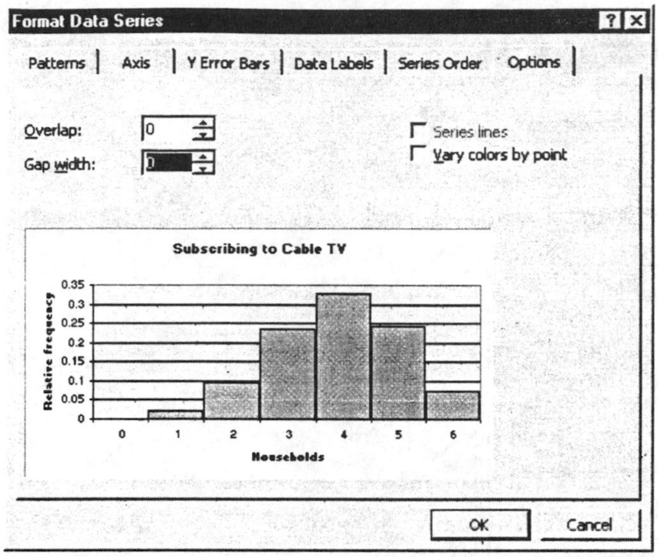

Section 4.3

▶ Example 3 (pg. 180) **Finding a Poisson Probability**

You will be using the POISSON function to find the probability that 2 rabbits are found on any given acre of a field when a population count shows that there is an average of 3.6 rabbits per acre living in the field.

1. Open a new Excel worksheet and click in cell **A1** to place the output there.

2. At the top of the screen, click **Insert** → **Function**.

3. Under Function category, select **Statistical**. Under Function name, select **POISSON**. Click **OK**.

Paste Function

Function category: Most Recently Used, All, Financial, Date & Time, Math & Trig, Statistical, Lookup & Reference, Database, Text, Logical, Information

Function name: MODE, NEGBINOMDIST, NORMDIST, NORMINV, NORMSDIST, NORMSINV, PEARSON, PERCENTILE, PERCENTRANK, PERMUT, POISSON

POISSON(x,mean,cumulative)

Returns the Poisson distribution.

OK Cancel

4. Complete the POISSON dialog box as shown below. Click **OK**.

POISSON

X 2 = 2

Mean 3.6 = 3.6

Cumulative FALSE = FALSE

= 0.177057721

Returns the Poisson distribution.

Cumulative is a logical value: for the cumulative Poisson probability, use TRUE; for the Poisson probability mass function, use FALSE.

Formula result =0.177057721 OK Cancel

The POISSON function returns as result of 0.177

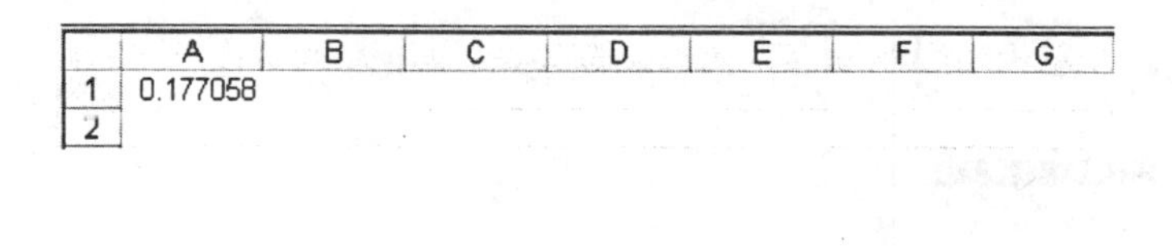

	A	B	C	D	E	F	G
1	0.177058						
2							

◀

Technology Lab

► Exercise 1 (pg. 185)

Creating a Poisson Distribution with μ = 4 for x = 0 to 10

1. Open a new Excel worksheet and enter the numbers 0 through 10 in column A as shown below. Then click in cell **B1** of the worksheet to place the output from the POISSON function there.

	A	B	C	D	E	F	G
1	0						
2	1						
3	2						
4	3						
5	4						
6	5						
7	6						
8	7						
9	8						
10	9						
11	10						

2. At the top of the screen, click **Insert → Function.**

3. Under Function category, select **Statistical**. Under Function name, select **POISSON**. Click **OK**.

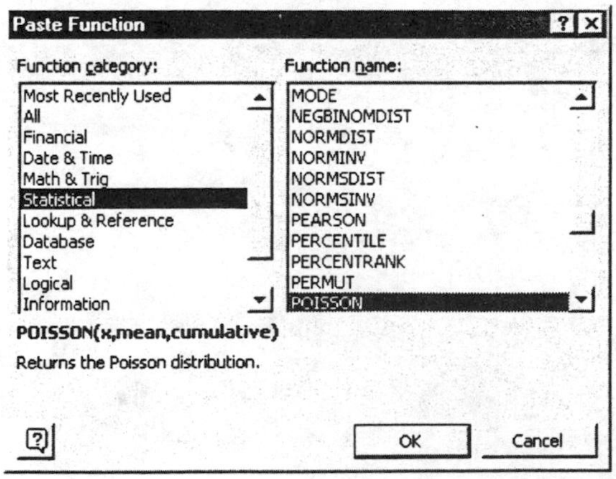

Paste Function

Function category:
Most Recently Used
All
Financial
Date & Time
Math & Trig
Statistical
Lookup & Reference
Database
Text
Logical
Information

Function name:
MODE
NEGBINOMDIST
NORMDIST
NORMINV
NORMSDIST
NORMSINV
PEARSON
PERCENTILE
PERCENTRANK
PERMUT
POISSON

POISSON(x,mean,cumulative)

Returns the Poisson distribution.

OK Cancel

4. Complete the POISSON dialog box as shown below. Click **OK**.

You are entering a relative cell address without dollar signs (i.e., A1) in the X field because you will be copying the contents of cell B1 to cells B2 through B11. You want the column A cell address to change from A1 to A2, A3, A4, ..., A11 when the formula is copied from cell B1 to cells B2 through B11.

POISSON

X A1 = 0

Mean 4 = 4

Cumulative FALSE = FALSE

= 0.018315639

Returns the Poisson distribution.

Cumulative is a logical value: for the cumulative Poisson probability, use TRUE; for the Poisson probability mass function, use FALSE.

Formula result =0.018315639 OK Cancel

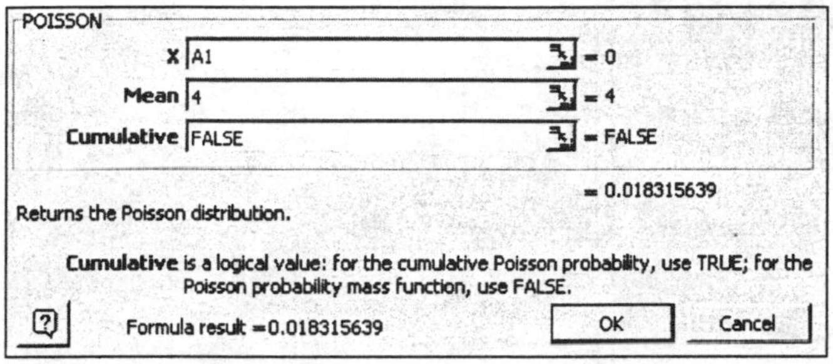

5. Copy the contents of cell B1 to cells B2 through B11.

	A	B	C	D	E	F	G
1	0	0.018316					
2	1	0.073263					
3	2	0.146525					
4	3	0.195367					
5	4	0.195367					
6	5	0.156293					
7	6	0.104196					
8	7	0.05954					
9	8	0.02977					
10	9	0.013231					
11	10	0.005292					

6. Construct a histogram of the Poisson distribution. First click on any cell of the table in the worksheet. Then, at the top of the screen, click **Insert → Chart**.

7. Under Chart type, in the Chart Type dialog box, select **Column**. Under Chart sub-type, select the first one in the top row by clicking on it. Click **Next**.

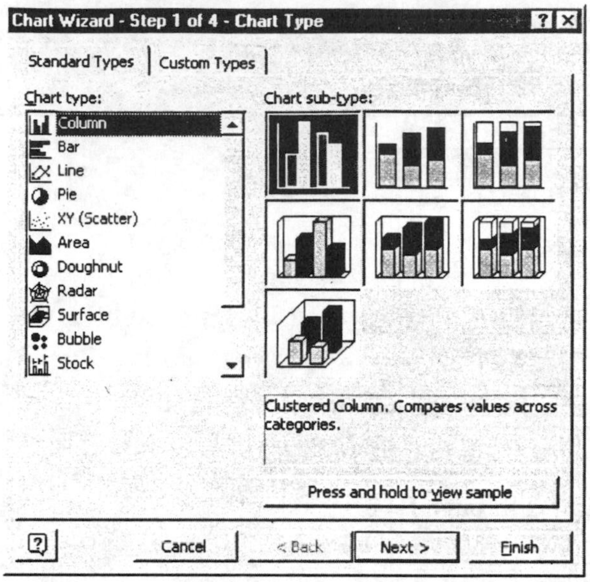

Chart Wizard - Step 1 of 4 - Chart Type

Standard Types | Custom Types

Chart type: Column, Bar, Line, Pie, XY (Scatter), Area, Doughnut, Radar, Surface, Bubble, Stock

Chart sub-type:

Clustered Column. Compares values across categories.

Press and hold to view sample

Cancel | < Back | Next > | Finish

8. Check the accuracy of the data range in the Chart Source Data dialog box. It should read **=Sheet1!$A$1:$B$11**. Make any necessary corrections. Then click **Next**.

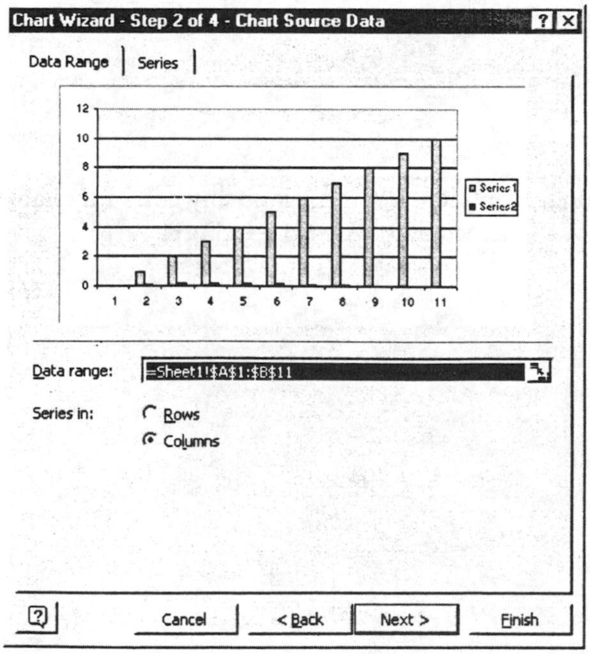

Chart Wizard - Step 2 of 4 - Chart Source Data

Data Range | Series

12
10
8
6
4
2
0
1 2 3 4 5 6 7 8 9 10 11
Series1
Series2

Data range: =Sheet1!$A$1:$B$11

Series in: Rows / Columns

Cancel | < Back | Next > | Finish

9. Click the **Titles** tab at the top of the Chart Options dialog box. In the Chart title window, enter **Customers Arriving at the Check-out Counter**. In the Category (X) axis window, enter **Number of Arrivals per Minute**. In the Value (Y) axis window, enter **Probability**.

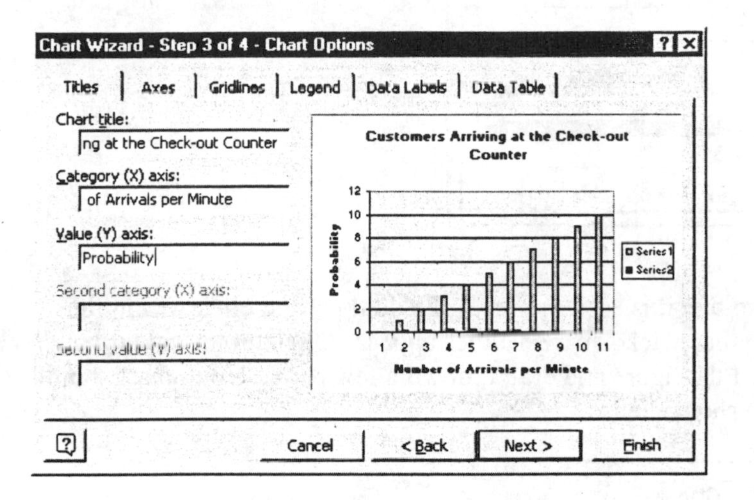

10. Click the **Legend** tab at the top of the Chart Options dialog box. Click in the box to the left of **Show Legend** to remove the checkmark. The removal of the checkmark will delete the Series 1 and Series 2 legends from the right side of the histogram chart. Click **OK**.

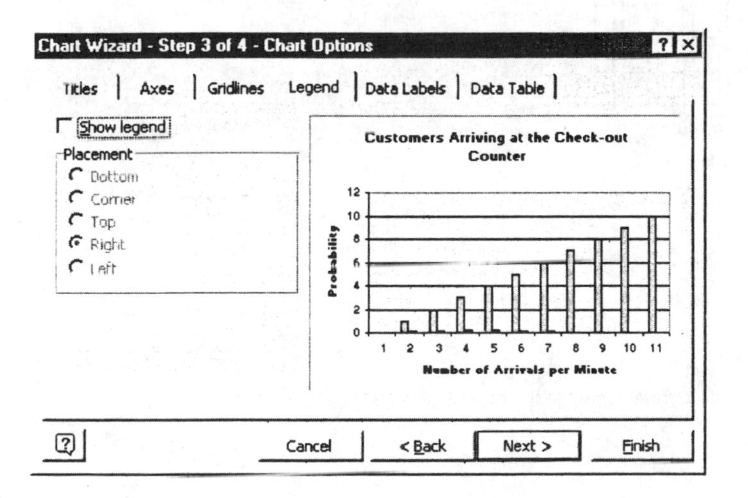

11. The Chart Location dialog box presents two options for placement of the chart. For this exercise, select **As object in**. Click **Finish**.

Chart Wizard - Step 4 of 4 - Chart Location

Place chart:

As new sheet: Chart1

As object in: Sheet1

Cancel | < Back | Next > | Finish

12. Make the chart taller so that it is easier to read. To do this, first click within the figure near a border so that black square handles appear. Click on the center handle on the bottom border of the figure and drag it down a few rows. Your chart should look similar to the one shown below.

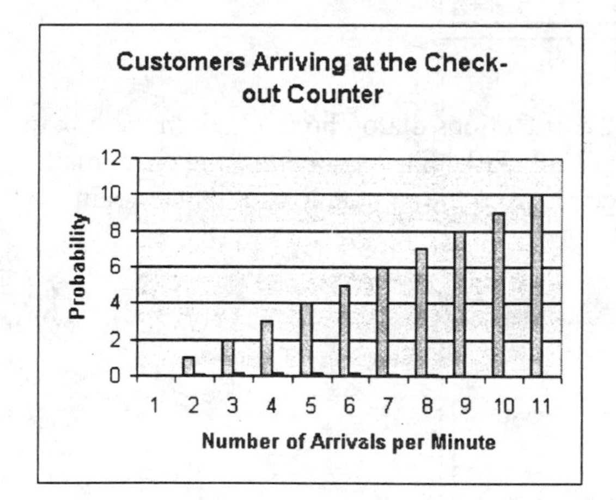

13. Correct the number scales displayed on the Y-axis and the X-axis. **Right click** on one of the vertical bars. Select **Source Data** from the shortcut menu that appears.

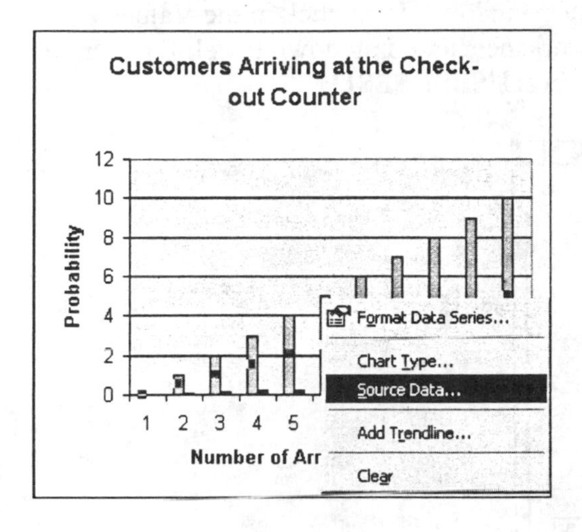

14. Click the **Series** tab at the top of the Source Data dialog box. In the Series window, select Series1 by clicking on it. Then click the **Remove** button below the Series window.

Source Data
Data Range
Series
Customers Arriving at the Check-out Counter
Probability
12
10
8
6
4
2
0
1 2 3 4 5 6 7 8 9 10 11
Number of Arrivals per Minute
Series
Series1
Series2
Name:
Values: =Sheet1!$A$1:$A$11
Add
Remove
Category (X) axis labels:
OK
Cancel

15. The probabilities that should be displayed on the Y-axis are located in column B of the worksheet. The entry in the Values window should be **=Sheet1!$B$1:$B$11**. First delete the information in the Values window. Then click in the Values window. Next click on cell B1 of the worksheet and drag down to cell B11. In the Values window you should now see **=Sheet1!$B$1:$B$11**.

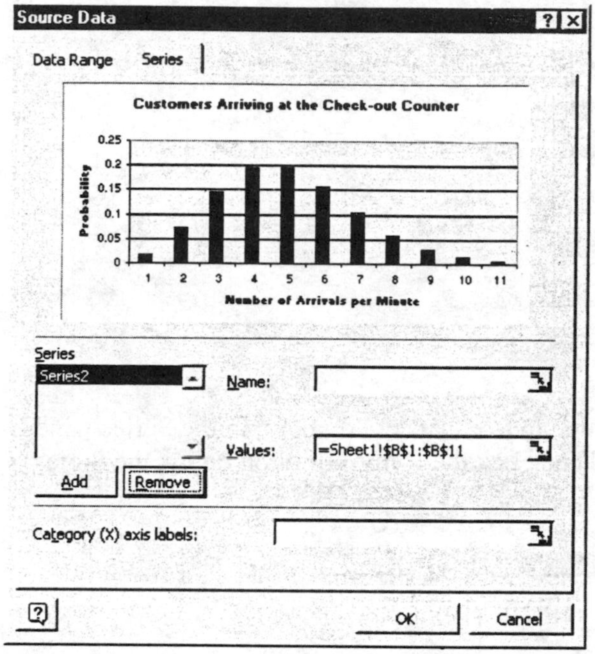

Source Data

Data Range Series

Customers Arriving at the Check-out Counter

Probability

0.25
0.2
0.15
0.1
0.05
0

1 2 3 4 5 6 7 8 9 10 11

Number of Arrivals per Minute

Series

Series2

Name:

Values: =Sheet1!$B$1:$B$11

Add Remove

Category (X) axis labels:

OK Cancel

16. The numbers you want displayed on the X-axis are in column A of the worksheet. Click in the Category (X) axis labels window. Then click on cell A1 of the worksheet and drag down to cell A11. The entry in the Category (X) axis labels field should now read **=Sheet1!$A$1:$A$11**. Click **OK**.

Source Data

Data Range Series

Customers Arriving at the Check-out Counter

Probability

0.25
0.2
0.15
0.1
0.05
0

0 1 2 3 4 5 6 7 8 9 10

Number of Arrivals per Minute

Series

Series2

Name:

Values: =Sheet1!$B$1:$B$11

Add Remove

Category (X) axis labels: =Sheet1!$A$1:$A$11

OK Cancel

17. Remove the space between the vertical bars. **Right click** on one of the vertical bars. Select **Format Data Series** from the shortcut menu that appears.

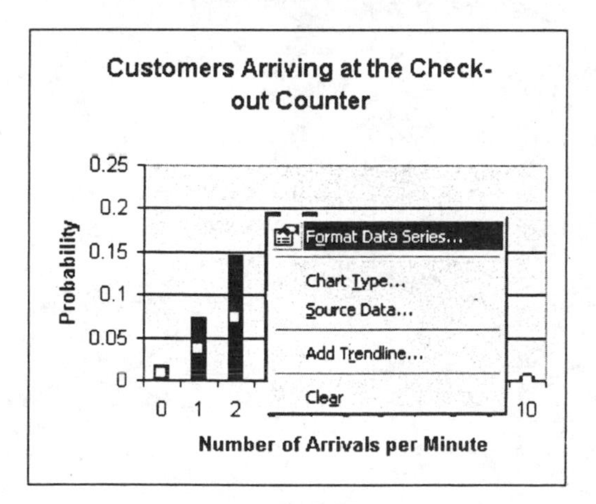

18. Click the **Options** tab at the top of the Format Data Series dialog box. Change the value in the Gap width box to 0. Click OK. Your histogram should look similar to the one shown below.

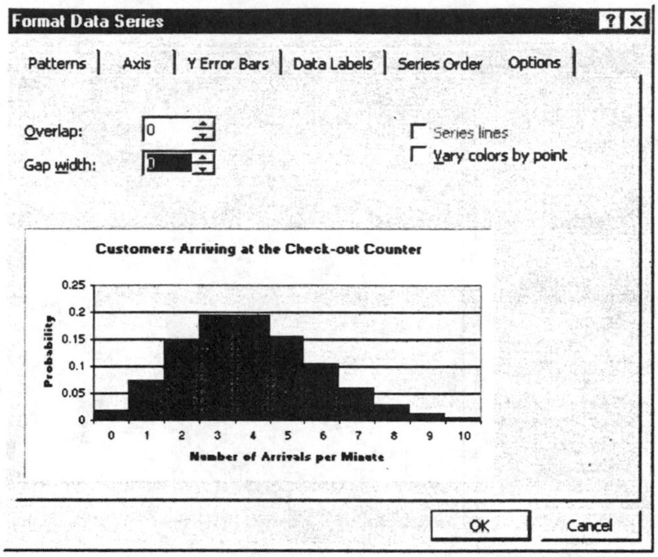

Format Data Series

Patterns | Axis | Y Error Bars | Data Labels | Series Order | Options

Overlap: 0

Gap width: 0

Series lines

Vary colors by point

Customers Arriving at the Check-out Counter

Probability

0.25
0.2
0.15
0.1
0.05
0

0 1 2 3 4 5 6 7 8 9 10

Number of Arrivals per Minute

OK Cancel

► Exercise 3 (pg. 185)

Generating a List of 20 Random Numbers with a Poisson distribution for μ = 4

1. Open a new, blank Excel worksheet. At the top of the screen, click **Tools → Data Analysis**. Select **Random Number Generation** and click **OK**.

If Data Analysis does not appear as a choice in the Tools menu, you will need to load the Microsoft Excel Analysis ToolPak add-in. Follow the procedure in Section GS 8.1 before continuing.

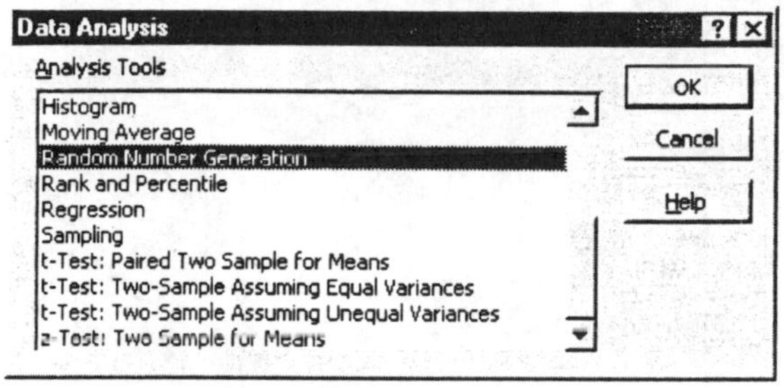

Data Analysis
Analysis Tools
Histogram
Moving Average
Random Number Generation
Rank and Percentile
Regression
Sampling
t-Test: Paired Two Sample for Means
t-Test: Two-Sample Assuming Equal Variances
t-Test: Two-Sample Assuming Unequal Variances
z-Test: Two Sample for Means
OK
Cancel
Help

2. Complete the Random Number Generation dialog box as shown below. The **Number of Variables** indicates the number of columns of values that you want in the output table. **Number of Random Numbers** you want generated is 20. Select the **Poisson** distribution. **Lambda** is the expected value (μ), which is equal to 4. The output will be placed in the current worksheet with A1 as the left topmost cell. Click **OK**.

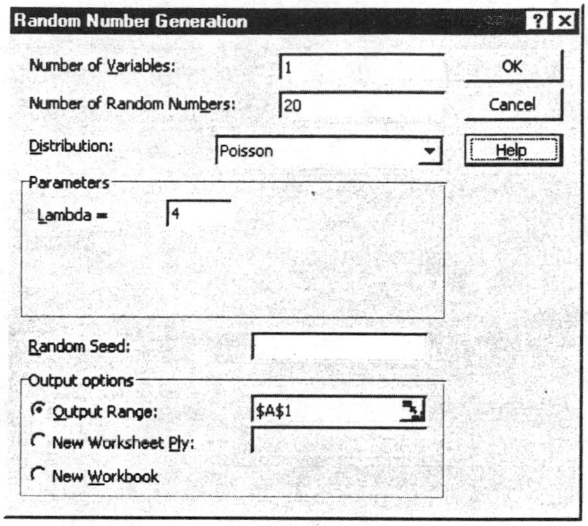

Random Number Generation

Number of Variables: 1
Number of Random Numbers: 20
Distribution: Poisson
Parameters
Lambda = 4
Random Seed:
Output options
Output Range: $A$1
New Worksheet Ply:
New Workbook
OK
Cancel
Help

The 20 numbers generated for this example are shown below. Because the numbers were generated randomly, it is not likely that your numbers will be exactly the same.

	A	B	C	D	E	F	G
1	3						
2	3						
3	4						
4	4						
5	5						
6	4						
7	5						
8	4						
9	5						
10	4						
11	3						
12	2						
13	4						
14	1						
15	6						
16	3						
17	7						
18	4						
19	3						
20	6						

◀

Normal Probability Distributions

Section 5.2

► Example 4 (pg. 207) Finding Area Under the Standard Normal Curve

1. Open a new Excel worksheet and click in cell **A1** to place the output there.

2. At the top of the screen, click **Insert → Function**.

3. Under Function category, select **Statistical**. Under Function name, select **NORMSDIST**. Click **OK**.

Paste Function

Function category:
Most Recently Used
All
Financial
Date & Time
Math & Trig
Statistical
Lookup & Reference
Database
Text
Logical
Information

Function name:
MINA
MODE
NEGBINOMDIST
NORMDIST
NORMINV
NORMSDIST
NORMSINV
PEARSON
PERCENTILE
PERCENTRANK
PERMUT

NORMSDIST(z)

Returns the standard normal cumulative distribution (has a mean of zero and a standard deviation of one).

OK Cancel

4. Complete the NORMSDIST dialog box as shown below. Click **OK**.

NORMSDIST

Z -.99 = -0.99

= 0.161087061

Returns the standard normal cumulative distribution (has a mean of zero and a standard deviation of one).

Z is the value for which you want the distribution.

Formula result =0.161087061 OK Cancel

The output is displayed in cell A1 of the worksheet. The area under the standard normal curve to the left of z = -0.99 is 0.161087.

	A	B	C	D	E	F	G
1	0.161087						
2							

◀

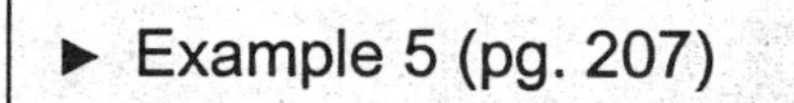

▶ Example 5 (pg. 207)

Finding Area Under the Standard Normal Curve

1. Open a new Excel worksheet. The area to the right of z = 1.06 is equal to 1 minus the area to the left of z = 1.06. In cell A1, enter **=1-** as shown below.

	A	B	C	D	E	F	G
1	=1-						
2							

2. At the top of the screen, click **Insert** → **Function**.

3. Under Function category, select **Statistical**. Under Function name, select **NORMSDIST**. Click **OK**.

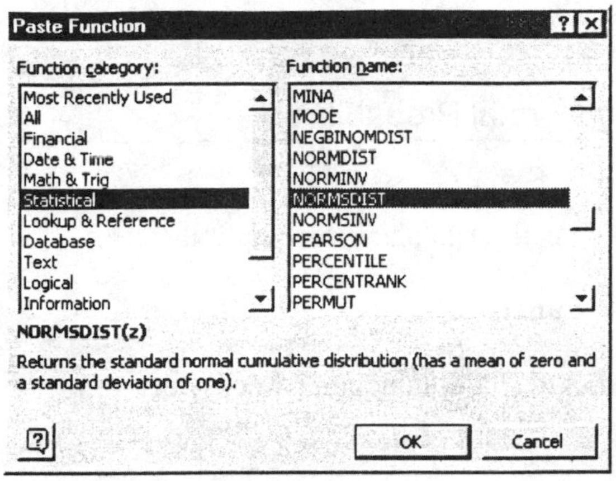

Paste Function

Function category:
Most Recently Used
All
Financial
Date & Time
Math & Trig
Statistical
Lookup & Reference
Database
Text
Logical
Information

Function name:
MINA
MODE
NEGBINOMDIST
NORMDIST
NORMINV
NORMSDIST
NORMSINV
PEARSON
PERCENTILE
PERCENTRANK
PERMUT

NORMSDIST(z)

Returns the standard normal cumulative distribution (has a mean of zero and a standard deviation of one).

OK Cancel

4. Complete the NORMSDIST dialog box as shown below. Click **OK**.

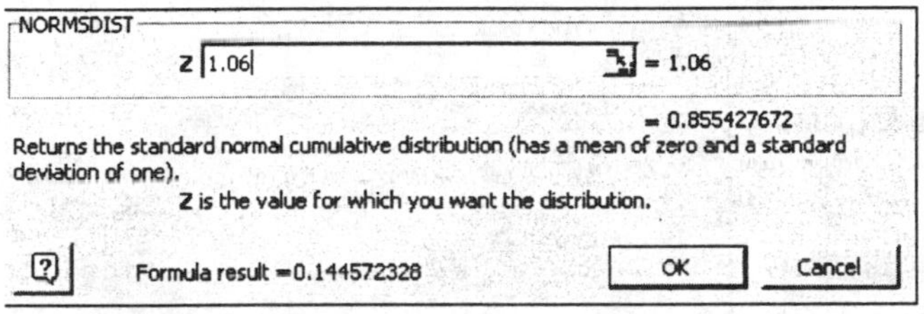

NORMSDIST

Z 1.06 = 1.06

= 0.855427672

Returns the standard normal cumulative distribution (has a mean of zero and a standard deviation of one).

Z is the value for which you want the distribution.

Formula result =0.144572328 OK Cancel

The output is displayed in cell A1 of the worksheet. The area under the standard normal curve to the right of z = 1.06 is 0.144572.

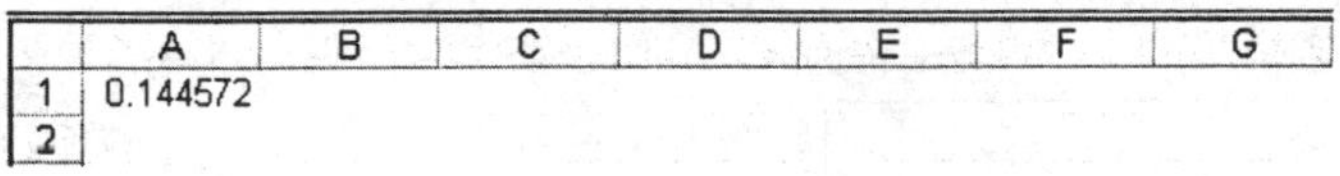

	A	B	C	D	E	F	G
1	0.144572						
2							

◀

Section 5.3

▶ Example 4 (pg. 215)

Finding Normal Probabilities

1. Open a new Excel worksheet and click in cell **A1** to place the output there.

2. At the top of the screen, click **Insert → Function**.

3. Under Function category, select **Statistical**. Under Function name, select **NORMDIST**. Click **OK**.

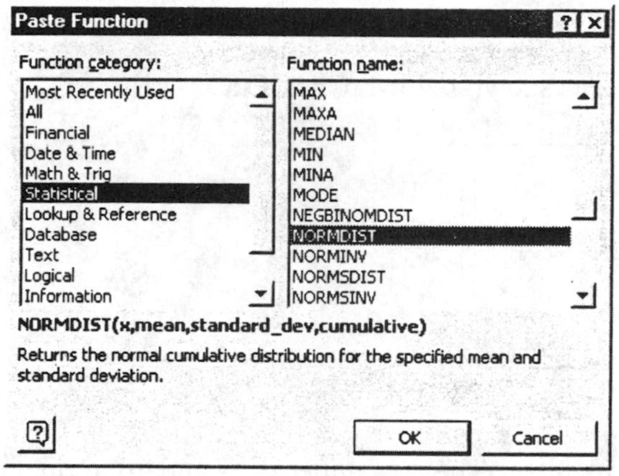

Paste Function

Function category: Most Recently Used, All, Financial, Date & Time, Math & Trig, Statistical, Lookup & Reference, Database, Text, Logical, Information

Function name: MAX, MAXA, MEDIAN, MIN, MINA, MODE, NEGBINOMDIST, NORMDIST, NORMINV, NORMSDIST, NORMSINV

NORMDIST(x,mean,standard_dev,cumulative)

Returns the normal cumulative distribution for the specified mean and standard deviation.

OK Cancel

4. Complete the NORMDIST dialog box as shown below. Click **OK**.

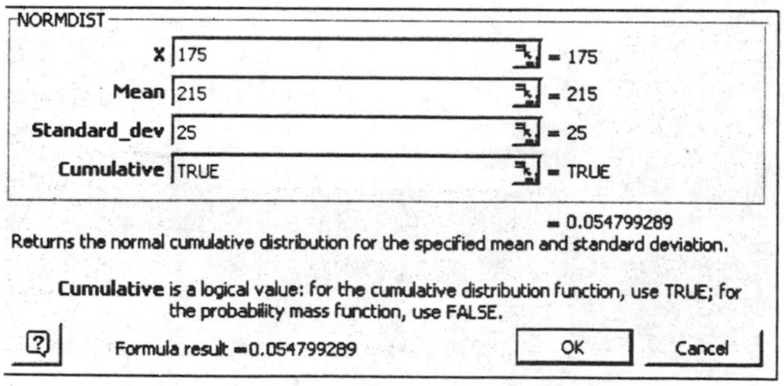

NORMDIST

X 175 = 175

Mean 215 = 215

Standard_dev 25 = 25

Cumulative TRUE = TRUE

= 0.054799289

Returns the normal cumulative distribution for the specified mean and standard deviation.

Cumulative is a logical value: for the cumulative distribution function, use TRUE; for the probability mass function, use FALSE.

Formula result =0.054799289 OK Cancel

The output is displayed in cell A1 of the worksheet. The probability is equal to 0.054799.

	A	B	C	D	E	F	G
1	0.054799						
2							

◀

Technology Lab

▶ Exercise 1 (pg. 234)

Finding the Mean Age in the United States

1. Open a new, blank Excel worksheet. Enter **Class Midpoint** in cell A1. You will be entering the numbers displayed in the table on page 234 of your text. Begin by entering the first two midpoints, **2.5** and **7.5**, as shown below.

	A	B	C	D	E	F	G
1	Class Midpoint						
2	2.5						
3	7.5						

2. You will now fill column A with a series. Click in cell A2 and drag down to cell A3 so that both cells are highlighted.

	A	B	C	D	E	F	G
1	Class Midpoint						
2	2.5						
3	7.5						

3. Move the mouse pointer in cell A3 to the right lower corner of the cell so that the white plus sign turns into a black plus sign. The black plus sign is called the "fill handle." Click the left mouse button and drag the fill handle down to cell A21.

	A	B	C	D	E	F	G
1	Class Midpoint						
2	2.5						
3	7.5						
4	12.5						
5	17.5						
6	22.5						
7	27.5						
8	32.5						
9	37.5						
10	42.5						
11	47.5						
12	52.5						
13	57.5						
14	62.5						
15	67.5						
16	72.5						
17	77.5						
18	82.5						
19	87.5						
20	92.5						
21	97.5						

4. Enter **Relative Frequency** in cell B1. Then enter the proportion equivalents of the percentages in column B as shown below.

	A	B	C	D	E	F	G
1	Class Midpoint	Relative Frequency					
2	2.5	0.073					
3	7.5	0.073					
4	12.5	0.072					
5	17.5	0.07					
6	22.5	0.066					
7	27.5	0.072					
8	32.5	0.081					
9	37.5	0.085					
10	42.5	0.078					
11	47.5	0.069					
12	52.5	0.052					
13	57.5	0.043					
14	62.5	0.038					
15	67.5	0.036					
16	72.5	0.035					
17	77.5	0.028					
18	82.5	0.015					
19	87.5	0.008					
20	92.5	0.005					
21	97.5	0.001					

5. Click in cell C2 and enter a formula to multiply the midpoint by the relative frequency. The formula is **=A2*B2**. Press **[Enter]**.

	A	B	C	D	E	F	G
1	Class Midpoint	Relative Frequency					
2	2.5	0.073	=A2*B2				

6. Copy the contents of cell C2 to cells C3 through C21.

	A	B	C	D	E	F	G
1	Class Midpoint	Relative Frequency					
2	2.5	0.073	0.1825				
3	7.5	0.073	0.5475				
4	12.5	0.072	0.9				
5	17.5	0.07	1.225				
6	22.5	0.066	1.485				
7	27.5	0.072	1.98				
8	32.5	0.081	2.6325				
9	37.5	0.085	3.1875				
10	42.5	0.078	3.315				
11	47.5	0.069	3.2775				
12	52.5	0.052	2.73				
13	57.5	0.043	2.4725				
14	62.5	0.038	2.375				
15	67.5	0.036	2.43				
16	72.5	0.035	2.5375				
17	77.5	0.028	2.17				
18	82.5	0.015	1.2375				
19	87.5	0.008	0.7				
20	92.5	0.005	0.4625				
21	97.5	0.001	0.0975				

7. The weighted mean is equal to the sum of the products in column C. (Refer to Section 2.3 in your text.) Click in cell **C22** of the worksheet to place the sum there. Click the AutoSum button near the top of the screen. It looks like this: Σ. The range of numbers to be included in the sum is displayed in cell C22. You should see **=SUM(C2:C21)**. Make any necessary corrections. Then press **[Enter]**.

	A	B	C	D	E	F	G
9	37.5	0.085	3.1875				
10	42.5	0.078	3.315				
11	47.5	0.069	3.2775				
12	52.5	0.052	2.73				
13	57.5	0.043	2.4725				
14	62.5	0.038	2.375				
15	67.5	0.036	2.43				
16	72.5	0.035	2.5375				
17	77.5	0.028	2.17				
18	82.5	0.015	1.2375				
19	87.5	0.008	0.7				
20	92.5	0.005	0.4625				
21	97.5	0.001	0.0975				
22			=SUM(C2:C21)				

8. The mean, displayed in cell C22, is 35.945 years. Save this worksheet so that you can use it again for Exercise 5, page 234.

◀

▶ Exercise 2 (pg. 234) Finding the Mean of the Set of Sample Means

1. Open worksheet "Tech5" in the Chapter 5 folder.

2. Click in cell **A38** at the bottom of the column of numbers to place the mean in that cell.

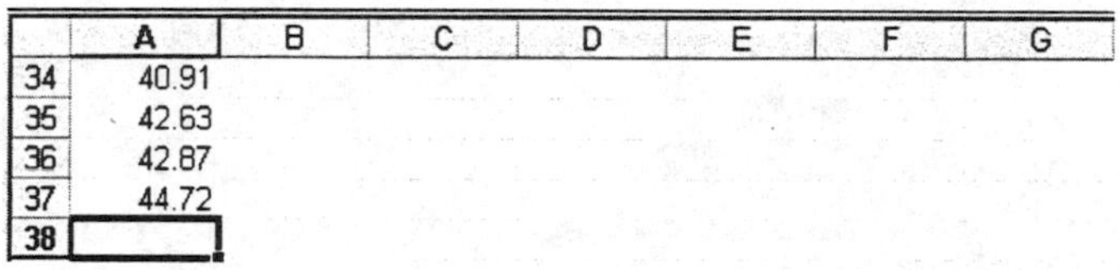

	A	B	C	D	E	F	G
34	40.91						
35	42.63						
36	42.87						
37	44.72						
38							

3. At the top of the screen, click **Insert → Function.**

4. Under Function category, select **Statistical**. Under Function name, select **AVERAGE**. Click **OK**.

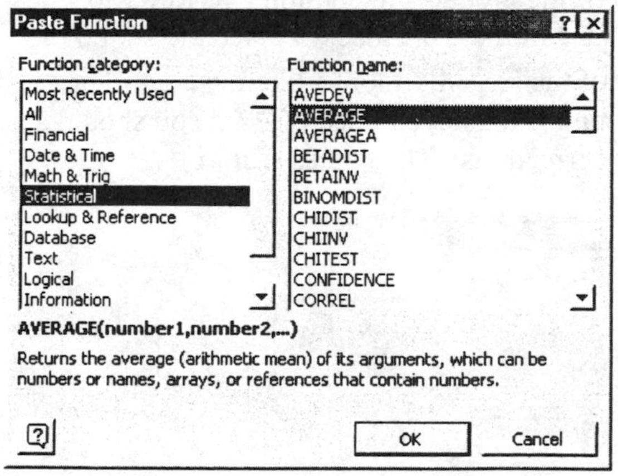

Paste Function

Function category:	Function name:
Most Recently Used	AVEDEV
All	AVERAGE
Financial	AVERAGEA
Date & Time	BETADIST
Math & Trig	BETAINV
Statistical	BINOMDIST
Lookup & Reference	CHIDIST
Database	CHIINV
Text	CHITEST
Logical	CONFIDENCE
Information	CORREL

AVERAGE(number1,number2,...)

Returns the average (arithmetic mean) of its arguments, which can be numbers or names, arrays, or references that contain numbers.

OK Cancel

5. The range that will be included in the average is shown in the Number 1 window. Check to be sure that it is accurate. It should read **A2:A37**. Make any necessary corrections. Then click **OK**.

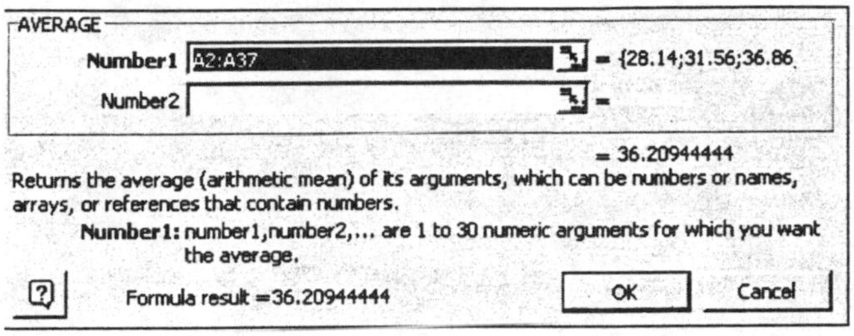

AVERAGE
Number1 A2:A37 = {28.14;31.56;36.86
Number2 =

= 36.20944444
Returns the average (arithmetic mean) of its arguments, which can be numbers or names, arrays, or references that contain numbers.
Number1: number1,number2,... are 1 to 30 numeric arguments for which you want the average.
Formula result =36.20944444 OK Cancel

The mean of the sample means is displayed in cell A38 of the worksheet. The mean is equal to 36.20944.

◀

▸ Exercise 5 (pg. 234) Finding the Standard Deviation of Ages in the United States

1. Open the worksheet that you prepared for Exercise 1, page 234.

	A	B	C	D	E	F	G
1	Class Midpoint	Relative Frequency					
2	2.5	0.073	0.1825				
3	7.5	0.073	0.5475				
4	12.5	0.072	0.9				
5	17.5	0.07	1.225				
6	22.5	0.066	1.485				
7	27.5	0.072	1.98				

2. To find the standard deviation of the population of ages, you will first calculate squared deviation scores. Click in cell D1 and enter the label, **Sqd Dev Score**.

	A	B	C	D	E	F	G
1	Class Midpoint	Relative Frequency		Sqd Dev Score			
2	2.5	0.073	0.1825				

3. Click in cell **D2** and enter the formula **=(A2-35.945)^2** and press **[Enter]**.

	A	B	C	D	E	F	G
1	Class Midpoint	Relative Frequency		Sqd Dev Score			
2	2.5	0.073	0.1825	=(A2-35.945)^2			

4. Copy the formula in cell D2 to cells D3 through D21.

	A	B	C	D	E	F	G
1	Class Midpoint	Relative Frequency		Sqd Dev Score			
2	2.5	0.073	0.1825	1118.568			
3	7.5	0.073	0.5475	809.118			
4	12.5	0.072	0.9	549.668			
5	17.5	0.07	1.225	340.218			
6	22.5	0.066	1.485	180.768			
7	27.5	0.072	1.98	71.31803			
8	32.5	0.081	2.6325	11.86803			
9	37.5	0.085	3.1875	2.418025			
10	42.5	0.078	3.315	42.96803			
11	47.5	0.069	3.2775	133.518			
12	52.5	0.052	2.73	274.068			
13	57.5	0.043	2.4725	464.618			
14	62.5	0.038	2.375	705.168			
15	67.5	0.036	2.43	995.718			
16	72.5	0.035	2.5375	1336.268			
17	77.5	0.028	2.17	1726.818			
18	82.5	0.015	1.2375	2167.368			
19	87.5	0.008	0.7	2657.918			
20	92.5	0.005	0.4625	3198.468			
21	97.5	0.001	0.0975	3789.018			

5. Each of the squared deviations will be weighted by its relative frequency. Click in cell E2. Enter the formula **=D2*B2** and press **[Enter]**.

	A	B	C	D	E	F	G
1	Class Midpoint	Relative Frequency		Sqd Dev Score			
2	2.5	0.073	0.1825	1118.568	=D2*B2		

6. Copy the formula in cell E2 to cells E3 through E21.

	A	B	C	D	E	F	G
1	Class Midpoint	Relative Frequency		Sqd Dev	Score		
2	2.5	0.073	0.1825	1118.568	81.65547		
3	7.5	0.073	0.5475	809.118	59.06562		
4	12.5	0.072	0.9	549.668	39.5761		
5	17.5	0.07	1.225	340.218	23.81526		
6	22.5	0.066	1.485	180.768	11.93069		
7	27.5	0.072	1.98	71.31803	5.134898		
8	32.5	0.081	2.6325	11.86803	0.96131		
9	37.5	0.085	3.1875	2.418025	0.205532		
10	42.5	0.078	3.315	42.96803	3.351506		
11	47.5	0.069	3.2775	133.518	9.212744		
12	52.5	0.052	2.73	274.068	14.25154		
13	57.5	0.043	2.4725	464.618	19.97858		
14	62.5	0.038	2.375	705.168	26.79638		
15	67.5	0.036	2.43	995.718	35.84585		
16	72.5	0.035	2.5375	1336.268	46.76938		
17	77.5	0.028	2.17	1726.818	48.3509		
18	82.5	0.015	1.2375	2167.368	32.51052		
19	87.5	0.008	0.7	2657.918	21.26334		
20	92.5	0.005	0.4625	3198.468	15.99234		
21	97.5	0.001	0.0975	3789.018	3.789018		

7. You are working with a relative frequency distribution of midpoints. For this type of distribution, the variance is equal to the sum of the squared deviation scores. Click in cell **E22** to place the sum of the squared deviation scores there. Click the AutoSum button near the top of the screen. It looks like this: Σ. The range of numbers to be included in the sum is displayed in cell E22. You should see **=SUM(E2:E21)**. Make any necessary corrections. Then press [**Enter**].

	A	B	C	D	E	F	G
18	82.5	0.015	1.2375	2167.368	32.51052		
19	87.5	0.008	0.7	2657.918	21.26334		
20	92.5	0.005	0.4625	3198.468	15.99234		
21	97.5	0.001	0.0975	3789.018	3.789018		
22			35.945		=SUM(E2:E21)		

8. Click in cell E23 to place the standard deviation there. The standard deviation is the square root of the variance. In cell E23, enter the formula **=sqrt(E22)** and press [**Enter**].

	A	B	C	D	E	F	G
18	82.5	0.015	1.2375	2167.368	32.51052		
19	87.5	0.008	0.7	2657.918	21.26334		
20	92.5	0.005	0.4625	3198.468	15.99234		
21	97.5	0.001	0.0975	3789.018	3.789018		
22			35.945		500.457		
23					=sqrt(E22)		

9. The standard deviation of ages in the United States is approximately equal to 22.3709. For future reference purposes, you may want to add labels for the mean, variance, and standard deviation as shown below.

	A	B	C	D	E	F	G
18	82.5	0.015	1.2375	2167.368	32.51052		
19	87.5	0.008	0.7	2657.918	21.26334		
20	92.5	0.005	0.4625	3198.468	15.99234		
21	97.5	0.001	0.0975	3789.018	3.789018		
22		Mean =	35.945	Var =	500.457		
23				St Dev =	22.3709		

▶ Exercise 6 (pg. 234)

Finding the Standard Deviation of the Set of Sample Means

1. Open worksheet "Tech5" in the Chapter 5 folder. This is the same data set that you used for Exercise 2, page 234. If you placed the mean (36.20944) in cell A38 for Exercise 2, page 234, you should delete it now.

	A	B	C	D	E	F	G
1	mean ages						
2	28.14						
3	31.56						
4	36.86						

2. At the top of the screen, click **Tools → Data Analysis.**

If Data Analysis does not appear as a choice in the Tools menu, you will need to load the Microsoft Excel Analysis ToolPak add-in. Follow the procedure in Section GS 8.1 before continuing.

---

3. Select **Descriptive Statistics** and click **OK**.

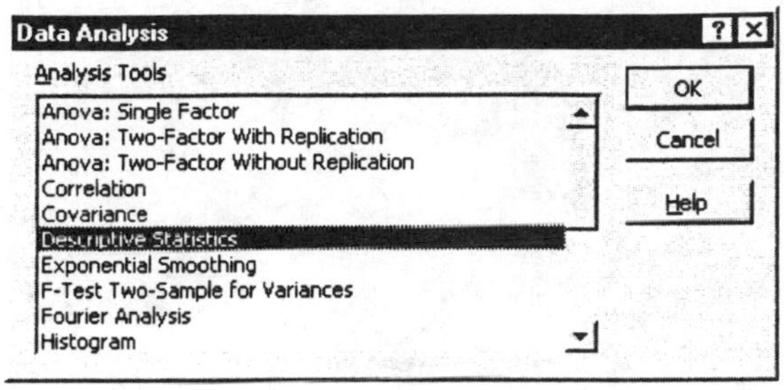

Data Analysis

Analysis Tools

Anova: Single Factor
Anova: Two-Factor With Replication
Anova: Two-Factor Without Replication
Correlation
Covariance
Descriptive Statistics
Exponential Smoothing
F-Test Two-Sample for Variances
Fourier Analysis
Histogram

OK
Cancel
Help

4. Complete the Descriptive Statistics dialog box as shown below. Click **OK**.

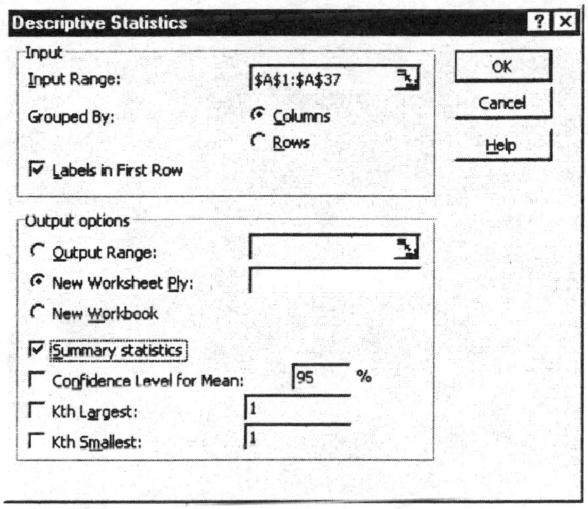

Descriptive Statistics

Input
Input Range: $A$1:$A$37
Grouped By: Columns / Rows
Labels in First Row

Output options
Output Range:
New Worksheet Ply:
New Workbook
Summary statistics
Confidence Level for Mean: 95 %
Kth Largest: 1
Kth Smallest: 1

OK
Cancel
Help

You will want to make column A wider so that you can read all the labels in the output table. The standard deviation is equal to 3.551804.

	A	B
1	*mean ages*	
2		
3	Mean	36.20944
4	Standard Error	0.591967
5	Median	36.155
6	Mode	#N/A
7	Standard Deviation	3.551804
8	Sample Variance	12.61531
9	Kurtosis	0.343186
10	Skewness	0.207283
11	Range	16.58
12	Minimum	28.14
13	Maximum	44.72
14	Sum	1303.54
15	Count	36

◄

Confidence Intervals

CHAPTER 6

Section 6.1

▶ Example 4 (pg. 256) — Constructing a Confidence Interval

If the PHStat add-in has not been loaded, you will need to load it before continuing. Follow the instructions in Section GS 8.2.

1. Open the "Sentence" worksheet in the Chapter 6 folder.

2. At the top of the screen, select **PHStat → Confidence Intervals → Estimate for the Mean, sigma known**.

3. Complete the Estimate for the Mean dialog box as shown below. Click **OK**.

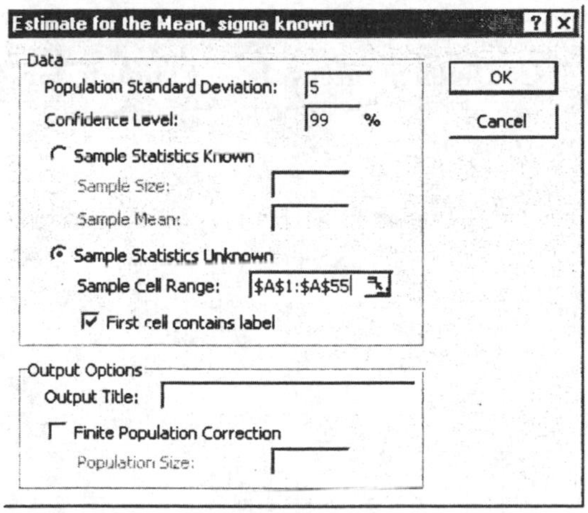

Estimate for the Mean, sigma known

Data
Population Standard Deviation: 5
Confidence Level: 99 %
Sample Statistics Known
Sample Size:
Sample Mean:
Sample Statistics Unknown
Sample Cell Range: $A$1:$A$55
First cell contains label
OK
Cancel

Output Options
Output Title:
Finite Population Correction
Population Size:

The output is displayed in a new worksheet. The lower limit of the confidence interval is 10.7 and the upper limit is 14.2.

	A	B
1	**Confidence Interval Estimate for the Mean**	
2		
3	**Population Standard Deviation**	**5**
4	**Sample Mean**	**12.42592593**
5	**Sample Size**	**54**
6	**Confidence Level**	**99%**
7	Standard Error of the Mean	0.680413817
8	Z Value	-2.57583451
9	Interval Half Width	1.752633395
10	**Interval Lower Limit**	**10.67329253**
11	**Interval Upper Limit**	**14.17855932**

◀

► Example 5 (pg. 257) Constructing a Confidence Interval, σ Known

If the PHStat add-in has not been loaded, you will need to load it before continuing. Follow the instructions in Section GS 8.2.

1. At the top of the screen, select **PHStat → Confidence Intervals → Estimate for the Mean, sigma known**.

2. Complete the Estimate for the Mean dialog box as shown below. Click **OK**.

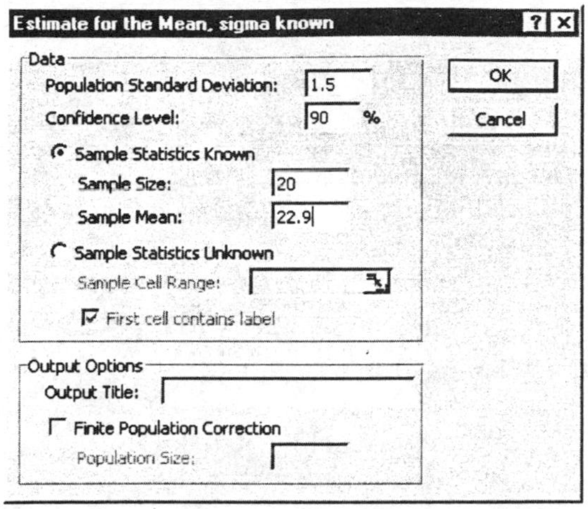

Estimate for the Mean, sigma known

Data
Population Standard Deviation: 1.5
Confidence Level: 90 %
Sample Statistics Known
Sample Size: 20
Sample Mean: 22.9
Sample Statistics Unknown
Sample Cell Range:
First cell contains label

Output Options
Output Title:
Finite Population Correction
Population Size:

OK
Cancel

Your output should look the same as the output displayed below. The lower limit of the 90% confidence interval is 22.3 and the upper limit is 23.5.

	A	B
1	**Confidence Interval Estimate for the Mean**	
2		
3	**Population Standard Deviation**	**1.5**
4	**Sample Mean**	**22.9**
5	**Sample Size**	**20**
6	**Confidence Level**	**90%**
7	Standard Error of the Mean	0.335410197
8	Z Value	-1.644853
9	Interval Half Width	0.551700468
10	**Interval Lower Limit**	**22.34829953**
11	**Interval Upper Limit**	**23.45170047**

Section 6.2

► Example 2 (pg. 268) Constructing a Confidence Interval, σ Unknown

If the PHStat add-in has not been loaded, you will need to load it before continuing. Follow the instructions in Section GS 8.2.

1. Open a new, blank Excel worksheet. At the top of the screen, select **PHStat → Confidence Intervals → Estimate for the Mean, sigma unknown**.

2. Complete the Estimate for the Mean dialog box as shown below. Click **OK**.

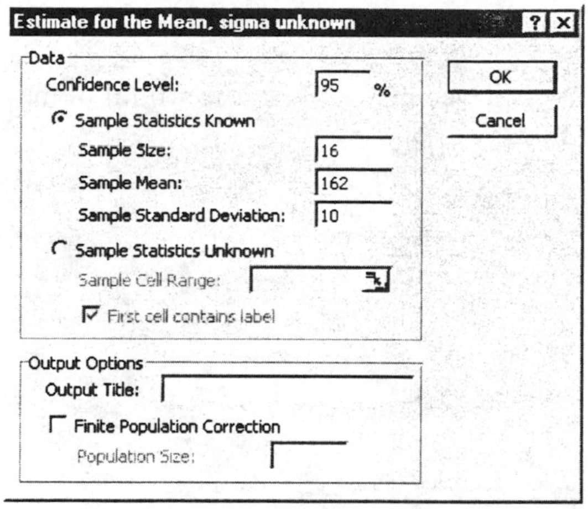

Estimate for the Mean, sigma unknown

Data
Confidence Level: 95 %
Sample Statistics Known
Sample Size: 16
Sample Mean: 162
Sample Standard Deviation: 10
Sample Statistics Unknown
Sample Cell Range:
First cell contains label

Output Options
Output Title:
Finite Population Correction
Population Size:

OK
Cancel

The output is displayed in a new worksheet. The lower limit of the 95% confidence interval is 156.67 and the upper limit is 167.33.

	A	B
1	**Confidence Interval Estimate for the Mean**	
2		
3	**Sample Standard Deviation**	**10**
4	**Sample Mean**	**162**
5	**Sample Size**	**16**
6	**Confidence Level**	**95%**
7	Standard Error of the Mean	2.5
8	Degrees of Freedom	15
9	*t* Value	2.131450856
10	Interval Half Width	5.32862714
11	**Interval Lower Limit**	**156.67**
12	**Interval Upper Limit**	**167.33**

◀

Section 6.3

▶ Example 2 (pg. 277) — Constructing a Confidence Interval for p

If the PHStat add-in has not been loaded, you will need to load it before continuing. Follow the instructions in Section GS 8.2.

1. Open a new, blank Excel worksheet. At the top of the screen, select **PHStat → Confidence Intervals → Estimate for the Proportion**.

2. Complete the Estimate for the Proportion dialog box as shown below. Click **OK**.

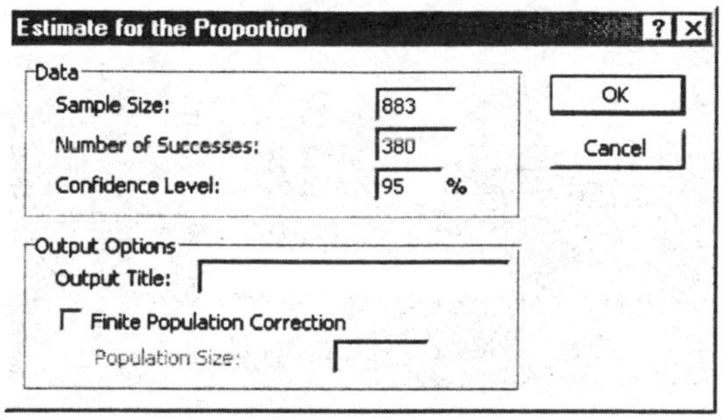

Estimate for the Proportion

Data
- Sample Size: 883
- Number of Successes: 380
- Confidence Level: 95 %

Output Options
- Output Title:
- ☐ Finite Population Correction
- Population Size:

OK | Cancel

Your output should look like the output displayed below. The lower limit of the 95% confidence interval is 0.397 and the upper limit is 0.463.

	A	B
1	**Confidence Interval Estimate for the Mean**	
2		
3	**Sample Size**	**883**
4	**Number of Successes**	**380**
5	**Confidence Level**	**95%**
6	Sample Proportion	0.430351076
7	Z Value	-1.95996108
8	Standard Error of the Proportion	0.016662292
9	Interval Half Width	0.032657443
10	**Interval Lower Limit**	**0.397693632**
11	**Interval Upper Limit**	**0.463008519**

◀

Technology Lab

► Exercise 1 (pg. 283) Finding a 95% Confidence Interval for p

If the PHStat add-in has not been loaded, you will need to load it before continuing. Follow the instructions in Section GS 8.2.

1. Open a new, blank Excel worksheet. At the top of the screen, select **PHStat → Confidence Intervals → Estimate for the Proportion.**

2. Complete the Estimate for the Proportion dialog box as shown below. Click **OK**.

Estimate for the Proportion ? X

Data
- Sample Size: 1005
- Number of Successes: 141
- Confidence Level: 95 %

OK | Cancel

Output Options
- Output Title:
- ☐ Finite Population Correction
- Population Size:

Your output should look the same as the output displayed below. The lower limit of the 95% confidence interval is 0.119 and the upper limit is 0.162.

	A	B
1	**Confidence Interval Estimate for the Mean**	
2		
3	**Sample Size**	**1005**
4	**Number of Successes**	**141**
5	**Confidence Level**	**95%**
6	Sample Proportion	0.140298507
7	Z Value	-1.95996108
8	Standard Error of the Proportion	0.010955125
9	Interval Half Width	0.021471619
10	**Interval Lower Limit**	**0.118826889**
11	**Interval Upper Limit**	**0.161770126**

► Exercise 3 (pg. 283) Finding a 95% confidence interval for p

If the PHStat add-in has not been loaded, you will need to load it before continuing. Follow the instructions in Section GS 8.2.

1. Open a new, blank Excel worksheet. At the top of the screen, select **PHStat → Confidence Intervals → Estimate for the Proportion.**

2. Six percent of the sample size, 1005, is 60. Use this number to complete the Estimate for the Proportion dialog box as shown below. Click **OK.**

Estimate for the Proportion

Data
- Sample Size: 1005
- Number of Successes: 60
- Confidence Level: 95 %

OK | Cancel

Output Options
- Output Title:
- Finite Population Correction
- Population Size:

Your output should look the same as the output displayed below. The upper limit of the 95% confidence interval is 0.045 and the upper limit is 0.074.

	A	B
1	**Confidence Interval Estimate for the Mean**	
2		
3	**Sample Size**	**1005**
4	**Number of Successes**	**60**
5	**Confidence Level**	**95%**
6	Sample Proportion	0.059701493
7	Z Value	-1.95996108
8	Standard Error of the Proportion	0.007473817
9	Interval Half Width	0.01464839
10	**Interval Lower Limit**	**0.045053103**
11	**Interval Upper Limit**	**0.074349882**

► Exercise 4 (pg. 283) Simulating a Most Admired Poll

1. Open a new, blank Excel worksheet. Enter the labels shown below for displaying the output of five simulations.

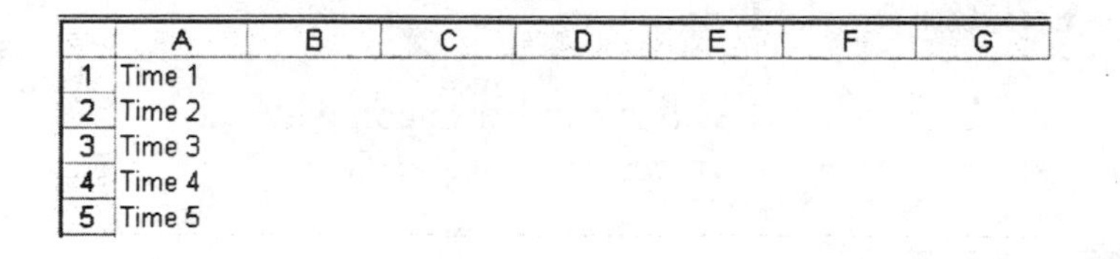

	A	B	C	D	E	F	G
1	Time 1						
2	Time 2						
3	Time 3						
4	Time 4						
5	Time 5						

2. At the top of the screen, select **Tools → Data Analysis**. Select **Random Number Generation**. Click **OK**.

If Data Analysis does not appear as a choice in the Tools menu, you will need to load the Microsoft Excel Analysis ToolPak add-in. Follow the procedure in Section GS 8.1 before continuing.

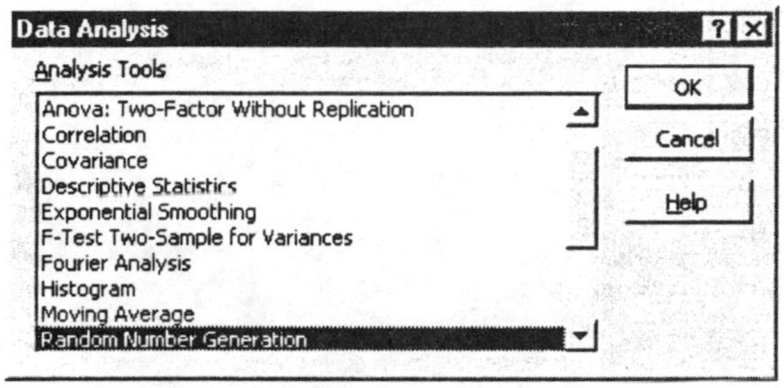

Data Analysis

Analysis Tools

Anova: Two-Factor Without Replication
Correlation
Covariance
Descriptive Statistics
Exponential Smoothing
F-Test Two-Sample for Variances
Fourier Analysis
Histogram
Moving Average
Random Number Generation

OK
Cancel
Help

3. Complete the Random Number Generation dialog box as shown below. Click **OK**.

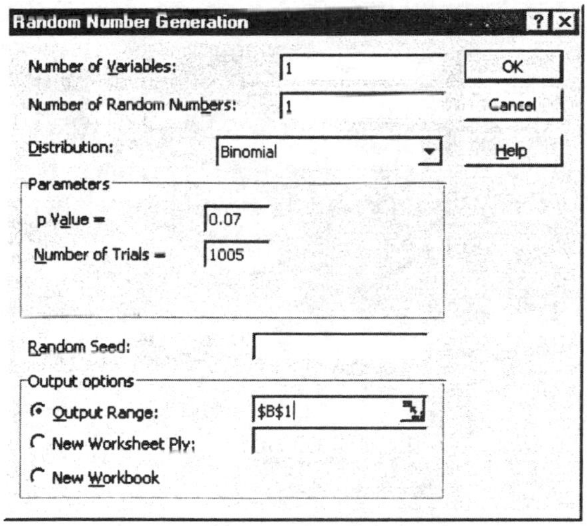

Random Number Generation

Number of Variables: 1
Number of Random Numbers: 1
Distribution: Binomial
Parameters
p Value = 0.07
Number of Trials = 1005
Random Seed:
Output options
Output Range: $B$1
New Worksheet Ply:
New Workbook

OK
Cancel
Help

For Time 1, the number of persons (out of 1005) selecting Oprah Winfrey is 84.

	A	B	C	D	E	F	G
1	Time 1	84					
2	Time 2						
3	Time 3						
4	Time 4						
5	Time 5						

4. Begin Time 2 the same way that you did Time 1. At the top of the screen, click **Tools → Data Analysis**. Select **Random Number Generation**. Click **OK**.

5. Complete the Random Number Generation dialog box in the same way as you did for Time 1 except change the location of the output to cell B2 as shown below. Click **OK**.

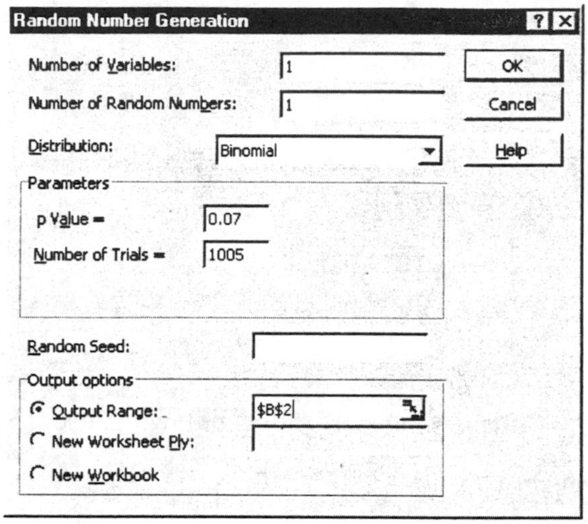

Random Number Generation

Number of Variables: 1

Number of Random Numbers: 1

Distribution: Binomial

Parameters

p Value = 0.07

Number of Trials = 1005

Random Seed:

Output options

Output Range: $B$2

New Worksheet Ply:

New Workbook

OK | Cancel | Help

6. Repeat this procedure until you have carried out the simulation five times. Your output should be similar to the output displayed below. Because the numbers were generated randomly, however, it is not likely that your numbers will be exactly the same as these.

	A	B	C	D	E	F	G
1	Time 1	84					
2	Time 2	66					
3	Time 3	71					
4	Time 4	76					
5	Time 5	77					

Hypothesis Testing with One Sample

Section 7.2

► Example 4 (pg. 318) — Testing μ with a Large Sample

If the PHStat add-in has not been loaded, you will need to load it before continuing. Follow the instructions in Section GS 8.2.

1. First open a new Excel worksheet. Then, at the top of the screen, select **PHStat → One-Sample Tests → Z Test for the Mean, sigma known**.

2. Complete the Z Test for the Mean dialog box as shown below. Click **OK**.

Z Test for the Mean, sigma known

Data
Null Hypothesis: 45000
Level of Significance: 0.05
Population Standard Deviation: 5200
Sample Statistics Known
Sample Size: 30
Sample Mean: 43500
Sample Statistics Unknown
Sample Cell Range:
First cell contains label

Test Options
Two-Tailed Test
Upper-Tail Test
Lower-Tail Test

Output Options
Output Title:

OK
Cancel

Your output should look like the output displayed below.

	A		B
1	**Z Test of Hypothesis for the Mean**		
2			
3	**Null Hypothesis**	**μ=**	**45000**
4	**Level of Significance**		**0.05**
5	**Population Standard Deviation**		**5200**
6	**Sample Size**		**30**
7	**Sample Mean**		**43500**
8	Standard Error of the Mean		949.3857663
9	Z Test Statistic		-1.579968916
10			
11	**Lower-Tail Test**		
12	**Lower Critical Value**		**-1.644853**
13	***p*-Value**		**0.057056996**
14	**Do not reject the null hypothesis**		

◀

▶ Example 9 (pg. 323)

Hypothesis Testing Using P-Values

If the PHStat add-in has not been loaded, you will need to load it before continuing. Follow the instructions in Section GS 8.2.

1. Open the "Franchise" worksheet in the Chapter 7 folder.
2. At the top of the screen, select **PHStat → One-Sample Tests → Z Test for the Mean, sigma known.**

3. Complete the Z Test for the Mean dialog box as shown below. Click **OK**.

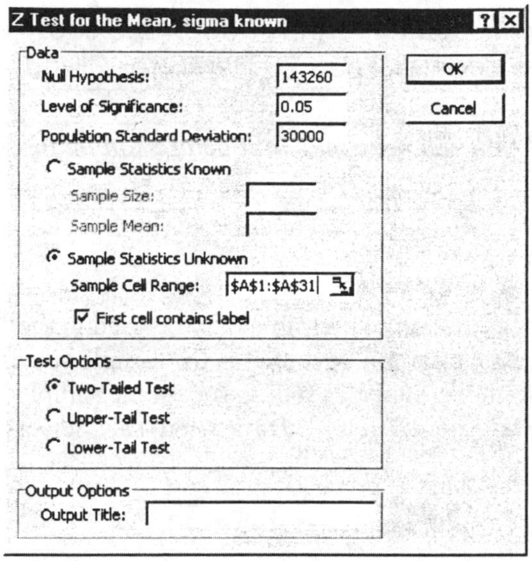

Z Test for the Mean, sigma known

Data
Null Hypothesis: 143260
Level of Significance: 0.05
Population Standard Deviation: 30000
Sample Statistics Known
Sample Size:
Sample Mean:
Sample Statistics Unknown
Sample Cell Range: $A$1:$A$31
First cell contains label

Test Options
Two-Tailed Test
Upper-Tail Test
Lower-Tail Test

Output Options
Output Title:

OK
Cancel

Your output should look like the output displayed below.

	A		B
1	**Z Test of Hypothesis for the Mean**		
2			
3	**Null Hypothesis**	μ=	**143260**
4	**Level of Significance**		**0.05**
5	**Population Standard Deviation**		**30000**
6	**Sample Size**		**30**
7	**Sample Mean**		**135000**
8	Standard Error of the Mean		5477.225575
9	Z Test Statistic		-1.508062775
10			
11	**Two-Tailed Test**		
12	**Lower Critical Value**		**-1.959961082**
13	**Upper Critical Value**		**1.959961082**
14	***p*-Value**		**0.131538512**
15	**Do not reject the null hypothesis**		

▶ Exercise 25 (pg. 326)

Testing That the Mean Nitrogen Dioxide Level Is Greater Than 28 Parts Per Billion

If the PHStat add-in has not been loaded, you will need to load it before continuing. Follow the instructions in Section GS 8.2.

1. Open the "Ex7_2-25" worksheet in the Chapter 7 folder.

2. Your textbook indicates that the standard deviation of a sample may be used in the z-test formula instead of a known population standard deviation if the sample size is 30 or greater. You will use Excel's Descriptive Statistics tool to obtain the sample standard deviation. At the top of the screen, click **Tools → Data Analysis**. Select **Descriptive Statistics**. Click **OK.**

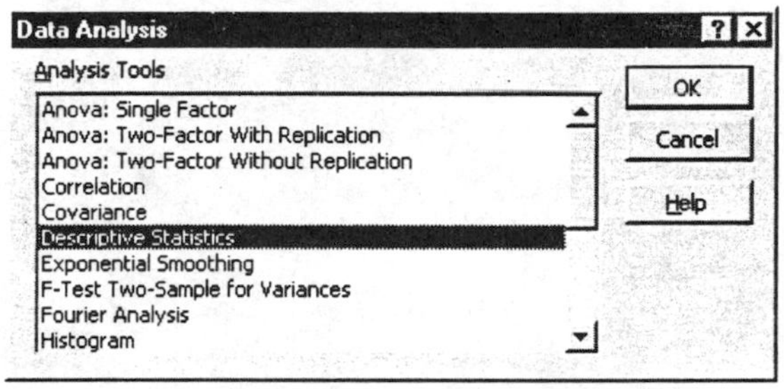

Data Analysis
Analysis Tools
Anova: Single Factor
Anova: Two-Factor With Replication
Anova: Two-Factor Without Replication
Correlation
Covariance
Descriptive Statistics
Exponential Smoothing
F-Test Two-Sample for Variances
Fourier Analysis
Histogram
OK
Cancel
Help

3. Complete the Descriptive Statistics dialog box as shown below. Click **OK**.

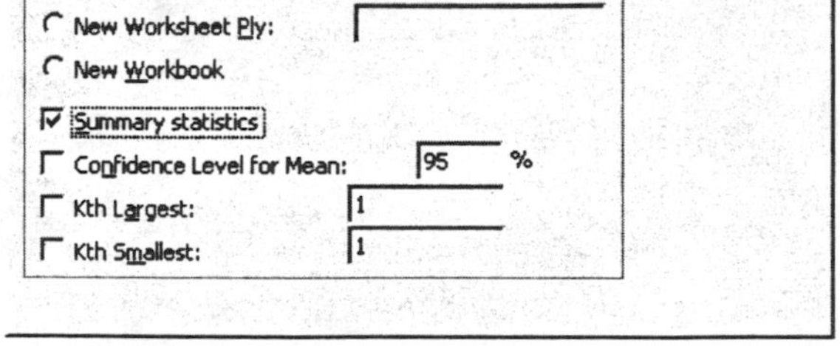

Descriptive Statistics
Input
Input Range: $A$1:$A$37
Grouped By: Columns
Rows
Labels in First Row
Output options
Output Range: $D$1
New Worksheet Ply:
New Workbook
Summary statistics
Confidence Level for Mean: 95 %
Kth Largest: 1
Kth Smallest: 1
OK
Cancel
Help

4. You will want to adjust the width of column D so that you can read all the labels. The standard deviation of the sample is 22.1284, and the mean is 32.8611. You will use this information when you carry out the statistical test. At the top of the screen, select **PHStat → One-Sample Tests → Z Test for the Mean, sigma known.**

	A	B	C	D	E
1	Nitrogen Dioxide levels			*Nitrogen Dioxide levels*	
2	27				
3	29			Mean	32.86111
4	53			Standard Error	3.688066
5	31			Median	28.5
6	16			Mode	29
7	47			Standard Deviation	22.1284
8	22			Sample Variance	489.6659
9	17			Kurtosis	2.150735
10	13			Skewness	1.497434
11	46			Range	90
12	99			Minimum	9
13	15			Maximum	99
14	20			Sum	1183
15	17			Count	36

5. Complete the Z Test for the Mean dialog box as shown below. Click **OK**.

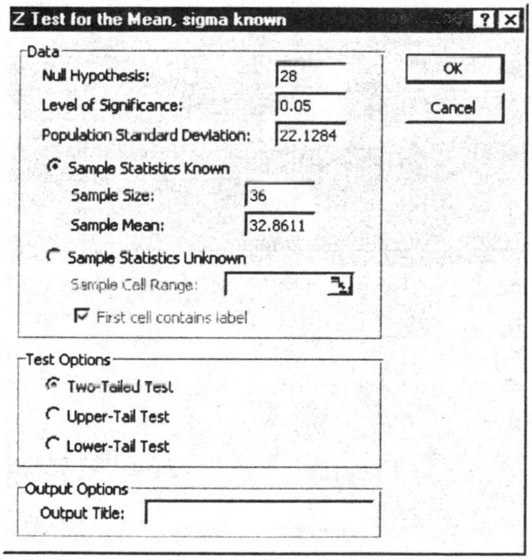

Z Test for the Mean, sigma known

Data
- Null Hypothesis: 28
- Level of Significance: 0.05
- Population Standard Deviation: 22.1284
- (•) Sample Statistics Known
  - Sample Size: 36
  - Sample Mean: 32.8611
- ( ) Sample Statistics Unknown
  - Sample Cell Range:
  - [x] First cell contains label

Test Options
- (•) Two-Tailed Test
- ( ) Upper-Tail Test
- ( ) Lower-Tail Test

Output Options
- Output Title:

OK | Cancel

Your output should look like the output displayed below.

	A	B
1	Z Test of Hypothesis for the Mean	
2		
3	Null Hypothesis $\mu=$	28
4	Level of Significance	0.05
5	Population Standard Deviation	22.1284
6	Sample Size	36
7	Sample Mean	32.8611
8	Standard Error of the Mean	3.688066667
9	Z Test Statistic	1.318061857
10		
11	Two-Tailed Test	
12	Lower Critical Value	-1.959961082
13	Upper Critical Value	1.959961082
14	p-Value	0.187483069
15	Do not reject the null hypothesis	

◀

Section 7.3

► Example 4 (pg. 333) Testing μ with a Small Sample

If the PHStat add-in has not been loaded, you will need to load it before continuing. Follow the instructions in Section GS 8.2

1. First open a new Excel worksheet. Then, at the top of the screen, select **PHStat → One-Sample Tests → t Test for the Mean, sigma unknown.**

2. Complete the t Test for the Mean dialog box as shown below. Click **OK**.

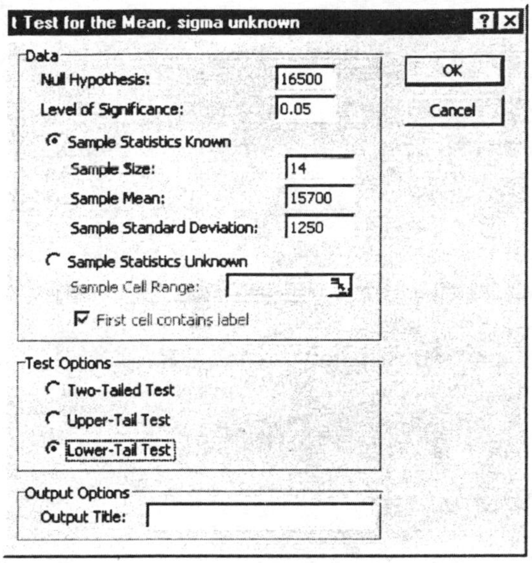

t Test for the Mean, sigma unknown

Data
Null Hypothesis: 16500
Level of Significance: 0.05
(•) Sample Statistics Known
Sample Size: 14
Sample Mean: 15700
Sample Standard Deviation: 1250
( ) Sample Statistics Unknown
Sample Cell Range:
[x] First cell contains label

Test Options
( ) Two-Tailed Test
( ) Upper-Tail Test
(•) Lower-Tail Test

Output Options
Output Title:

OK
Cancel

Your output should look like the output shown below.

	A		B	C	D	E
1	t Test of Hypothesis for the Mean					
2						
3	Null Hypothesis	μ=	16500			
4	Level of Significance		0.05			
5	Sample Size		14			
6	Sample Mean		15700			
7	Sample Standard Deviation		1250			
8	Standard Error of the Mean		334.0765524			
9	Degrees of Freedom		13			
10	*t* Test Statistic		-2.394660728			
11						
12	Lower-Tail Test					
13	Lower Critical Value		-1.770931704			
14	*p*-Value		0.016203608			
15	Reject the null hypothesis					
16						
17					Calculations Area	
18					For one-tailed tests:	
19					TDIST value	0.016204
20					1-TDIST value	0.983796

◀

► Example 6 (pg. 335) Using P-Values with a t-Test

If the PHStat add-in has not been loaded, you will need to load it before continuing. Follow the instructions in Section GS 8.2.

1. First open a new Excel worksheet. Then, at the top of the screen, select **PHStat → One-Sample Tests → t Test for the Mean, sigma unknown**.

2. Complete the t Test for the Mean dialog box as shown below. Click **OK**.

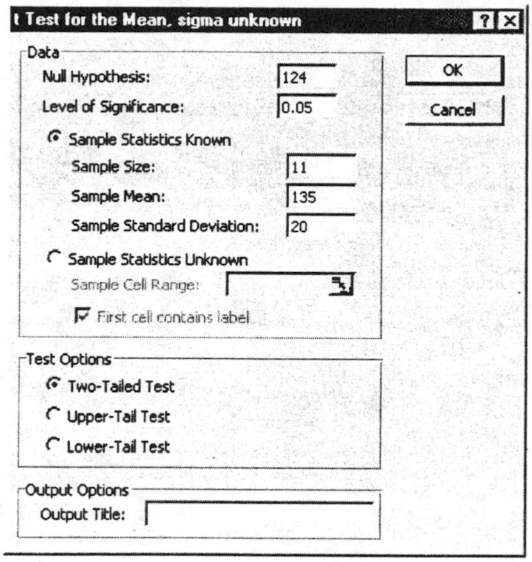

t Test for the Mean, sigma unknown

Data
Null Hypothesis: 124
Level of Significance: 0.05
Sample Statistics Known
Sample Size: 11
Sample Mean: 135
Sample Standard Deviation: 20
Sample Statistics Unknown
Sample Cell Range:
First cell contains label

Test Options
Two-Tailed Test
Upper-Tail Test
Lower-Tail Test

Output Options
Output Title:

OK
Cancel

Your output should look like the output displayed below.

	A	B	C	D	E
1	t Test of Hypothesis for the Mean				
2					
3	Null Hypothesis μ=	124			
4	Level of Significance	0.05			
5	Sample Size	11			
6	Sample Mean	135			
7	Sample Standard Deviation	20			
8	Standard Error of the Mean	6.030226892			
9	Degrees of Freedom	10			
10	t Test Statistic	1.824143635			
11					
12	Two-Tailed Test				
13	Lower Critical Value	-2.228139238			
14	Upper Critical Value	2.228139238			
15	p-Value	0.098110528			
16	Do not reject the null hypothesis				
17				Calculations Area	
18				For one-tailed tests:	
19				TDIST value	0.049055
20				1-TDIST value	0.950945

Testing the Claim That the Mean Recycled Waste Is More Than 1 Lb. Per Day

If the PHStat add-in has not been loaded, you will need to load it before continuing. Follow the instructions in Section GS 8.2.

1. First open a new Excel worksheet. Then, at the top of the screen, select **PHStat → One-Sample Tests → t Test for the Mean, sigma unknown**.

2. Complete the t Test for the Mean dialog box as shown below. Click **OK**.

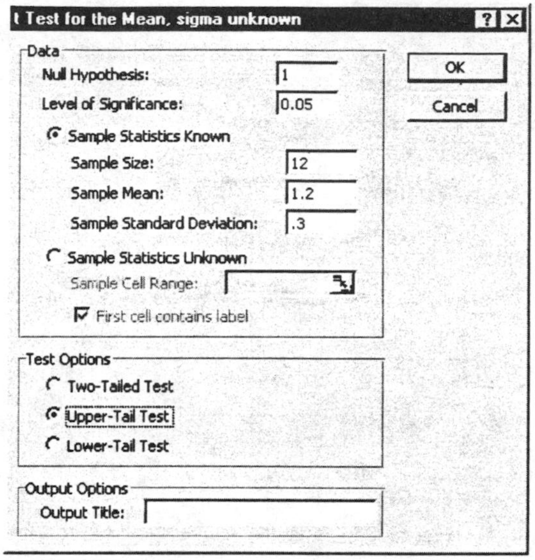

t Test for the Mean, sigma unknown

Data
- Null Hypothesis: 1
- Level of Significance: 0.05
- (•) Sample Statistics Known
  - Sample Size: 12
  - Sample Mean: 1.2
  - Sample Standard Deviation: .3
- ( ) Sample Statistics Unknown
  - Sample Cell Range:
  - [x] First cell contains label

Test Options
- ( ) Two-Tailed Test
- (•) Upper-Tail Test
- ( ) Lower-Tail Test

Output Options
- Output Title:

OK | Cancel

Your output should be the same as the output displayed below.

	A		B	C	D	E
1	**t Test of Hypothesis for the Mean**					
2						
3	**Null Hypothesis**	**μ=**	**1**			
4	**Level of Significance**		**0.05**			
5	**Sample Size**		**12**			
6	**Sample Mean**		**1.2**			
7	**Sample Standard Deviation**		**0.3**			
8	Standard Error of the Mean		0.08660254			
9	Degrees of Freedom		11			
10	*t* Test Statistic		2.309401077			
11						
12	**Upper-Tail Test**					
13	**Upper Critical Value**		**1.795883691**			
14	***p*-Value**		**0.020671133**			
15	**Reject the null hypothesis**					
16						
17					Calculations Area	
18					For one-tailed tests:	
19					TDIST value	0.020671
20					1-TDIST value	0.979329

◀

▶ Exercise 23 (pg. 338)

Testing the Claim That the Mean Number of 12-Oz. Servings Is Less Than 3.0

If the PHStat add-in has not been loaded, you will need to load it before continuing. Follow the instructions in Section GS 8.2.

1. Open the "Ex7_3-23" worksheet in the Chapter 7 folder.

2. At the top of the screen, select **PHStat → One-Sample Tests → t Test for the Mean, sigma unknown.**

3. Complete the t Test for the Mean dialog box as shown below. Click **OK.**

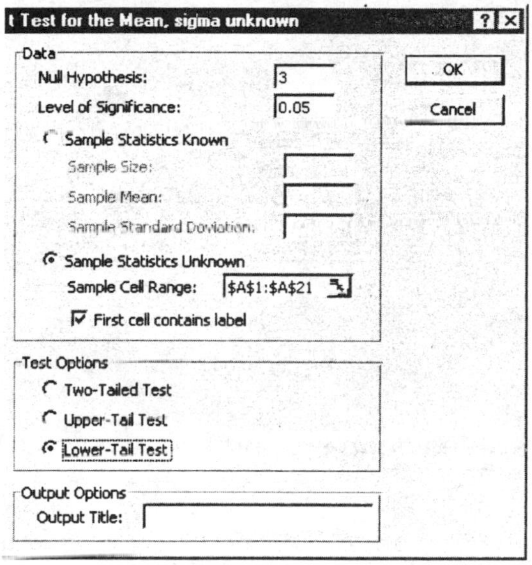

t Test for the Mean, sigma unknown

Data
Null Hypothesis: 3
Level of Significance: 0.05
Sample Statistics Known
Sample Size:
Sample Mean:
Sample Standard Deviation:
Sample Statistics Unknown
Sample Cell Range: $A$1:$A$21
First cell contains label

Test Options
Two-Tailed Test
Upper-Tail Test
Lower-Tail Test

Output Options
Output Title:

OK
Cancel

Your output should look like the output displayed below.

	A	B	C	D	E
1	**t Test of Hypothesis for the Mean**				
2					
3	**Null Hypothesis** μ=	**3**			
4	**Level of Significance**	**0.05**			
5	**Sample Size**	**20**			
6	**Sample Mean**	**2.785**			
7	**Sample Standard Deviation**	**0.827981058**			
8	Standard Error of the Mean	0.185142193			
9	Degrees of Freedom	19			
10	*t* Test Statistic	-1.161269599			
11					
12	**Lower-Tail Test**				
13	**Lower Critical Value**	**-1.729131327**			
14	***p*-Value**	**0.129957182**			
15	**Do not reject the null hypothesis**				
16					
17				Calculations Area	
18				For one-tailed tests:	
19				TDIST value	0.129957
20				1-TDIST value	0.870043

◀

Section 7.4

► Example 1 (pg. 341) **Hypothesis Test for a Proportion**

If the PHStat add-in has not been loaded, you will need to load it before continuing. Follow the instructions in Section GS 8.2.

1. First open a new Excel worksheet. Then, at the top of the screen, select **PHStat → One-Sample Tests → Z Test for the Proportion**.

2. Complete the Z Test for the Proportion dialog box as shown below. Click **OK**.

Z Test for the Proportion

Data
- Null Hypothesis: .2
- Level of Significance: 0.05
- Number of Successes: 15
- Sample Size: 100

OK | Cancel

Test Options
- ( ) Two-Tailed Test
- ( ) Upper-Tail Test
- (•) Lower-Tail Test

Output Options
- Output Title:

Your output should be the same as the output displayed below.

	A		B
1	**Z Test of Hypothesis for the Proportion**		
2			
3	**Null Hypothesis**	*p*=	**0.2**
4	**Level of Significance**		**0.05**
5	**Number of Successes**		**15**
6	**Sample Size**		**100**
7	Sample Proportion		0.15
8	Standard Error		0.04
9	Z Test Statistic		-1.25
10			
11	**Lower-Tail Test**		
12	**Lower Critical Value**		**-1.644853**
13	***p*-Value**		**0.105649839**
14	**Do not reject the null hypothesis**		

◀

▶ Exercise 9 (pg. 344)

Testing the Claim That More Than 30% of Consumers Have Stopped Buying a Product

If the PHStat add-in has not been loaded, you will need to load it before continuing. Follow the instructions in Section GS 8.2.

1. First open a new Excel worksheet. Then, at the top of the screen, select **PHStat → One-Sample Tests → Z Test for the Proportion**.

2. Complete the Z Test for the Proportion dialog box as shown below. Click **OK**.

Note that 32% of 1050 is 336.

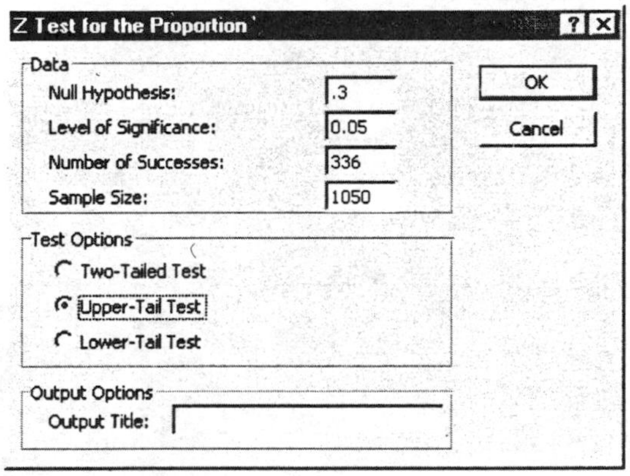

Z Test for the Proportion

Data
Null Hypothesis: .3
Level of Significance: 0.05
Number of Successes: 336
Sample Size: 1050

OK
Cancel

Test Options
Two-Tailed Test
Upper-Tail Test
Lower-Tail Test

Output Options
Output Title:

Your output should be the same as the output displayed below.

	A	B	C
1	**Z Test of Hypothesis for the Proportion**		
2			
3	**Null Hypothesis** $p=$	**0.3**	
4	**Level of Significance**	**0.05**	
5	**Number of Successes**	**336**	
6	**Sample Size**	**1050**	
7	Sample Proportion	0.32	
8	Standard Error	0.014142136	
9	Z Test Statistic	1.414213562	
10			
11	**Upper-Tail Test**		
12	**Upper Critical Value**	**1.644853**	
13	***p*-Value**	**0.078649653**	
14	**Do not reject the null hypothesis**		

◄

► Technology Lab (pg. 346) Testing the Claim That the Proportion Is Equal to 0.2914

If the PHStat add-in has not been loaded, you will need to load it before continuing. Follow the instructions in Section GS 8.2.

1. First open a new Excel worksheet. Then, at the top of the screen, select **PHStat → One-Sample Tests → Z Test for the Proportion**.

2. Complete the Z Test for the Proportion dialog box as shown below. Click **OK**.

Z Test for the Proportion

Data
- Null Hypothesis: .2914
- Level of Significance: 0.05
- Number of Successes: 9
- Sample Size: 100

Test Options
- (•) Two-Tailed Test
- ( ) Upper-Tail Test
- ( ) Lower-Tail Test

Output Options
- Output Title:

OK | Cancel

Your output should appear the same as the output shown below.

	A	B
1	**Z Test of Hypothesis for the Proportion**	
2		
3	**Null Hypothesis** *p*=	**0.2914**
4	**Level of Significance**	**0.05**
5	**Number of Successes**	**9**
6	**Sample Size**	**100**
7	Sample Proportion	0.09
8	Standard Error	0.045440735
9	Z Test Statistic	-4.432146616
10		
11	**Two-Tailed Test**	
12	**Lower Critical Value**	**-1.959961082**
13	**Upper Critical value**	**1.959961082**
14	***p*-Value**	**9.33787E-06**
15	**Reject the null hypothesis**	

◄

Section 7.5

► Exercise 25 (pg. 355)

Using the P-Value Method to Perform the Hypothesis Test for Exercise 23

1. Open a new Excel worksheet and click in cell **A1** to place the output there.

2. At the top of the screen, select **Insert → Function**.

3. Under Function category, select **Statistical**. Under Function name, select **CHIDIST**. Click **OK**.

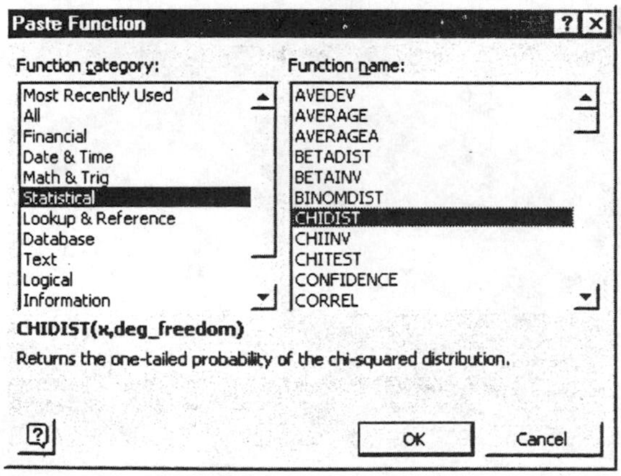

Paste Function

Function category: Most Recently Used, All, Financial, Date & Time, Math & Trig, Statistical, Lookup & Reference, Database, Text, Logical, Information

Function name: AVEDEV, AVERAGE, AVERAGEA, BETADIST, BETAINV, BINOMDIST, CHIDIST, CHIINV, CHITEST, CONFIDENCE, CORREL

CHIDIST(x,deg_freedom)

Returns the one-tailed probability of the chi-squared distribution.

OK Cancel

4. Complete the CHIDIST dialog box as shown below. Click **OK**.

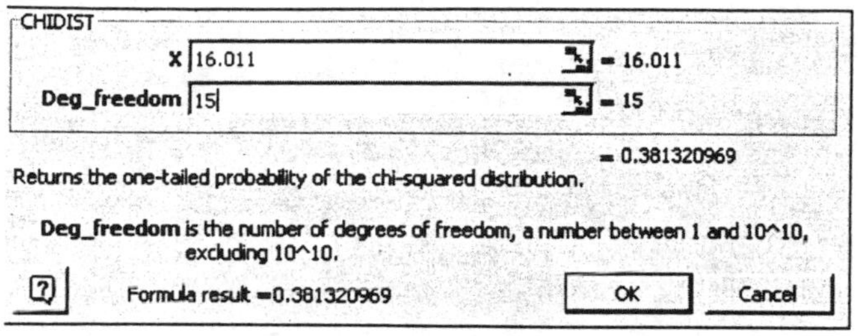

CHIDIST

X 16.011 = 16.011

Deg_freedom 15 = 15

= 0.381320969

Returns the one-tailed probability of the chi-squared distribution.

Deg_freedom is the number of degrees of freedom, a number between 1 and 10^10, excluding 10^10.

Formula result =0.381320969 OK Cancel

With df=15, the one-tailed probability of X^2 is 0.3813.

	A	B	C	D	E	F	G
1	0.381321						
2							

Hypothesis Testing with Two Samples

CHAPTER 8

Section 8.1

► Exercise 13 (pg. 375)

Testing the Claim That Children Ages 3-12 Watched TV More in 1981 Than Today

1. Open worksheet "Ex8_1-13" in the Chapter 8 folder.

2. Your textbook indicates that the variance of a sample may be used in the z-test formula instead of a known population variance if the sample size is sufficiently large. You will use Excel's Descriptive Statistics tool to obtain the sample variance. At the top of the screen, click **Tools → Data Analysis**. Select **Descriptive Statistics**. Click **OK**.

If Data Analysis does not appear as a choice in the Tools menu, you will need to load the Microsoft Excel ToolPak add-in. Follow the procedure in Section GS 8.1 before continuing.

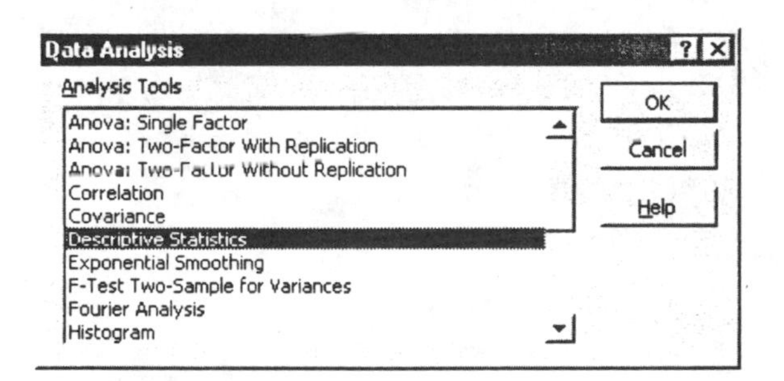

3. Complete the Descriptive Statistics dialog box as shown below. Click **OK**.

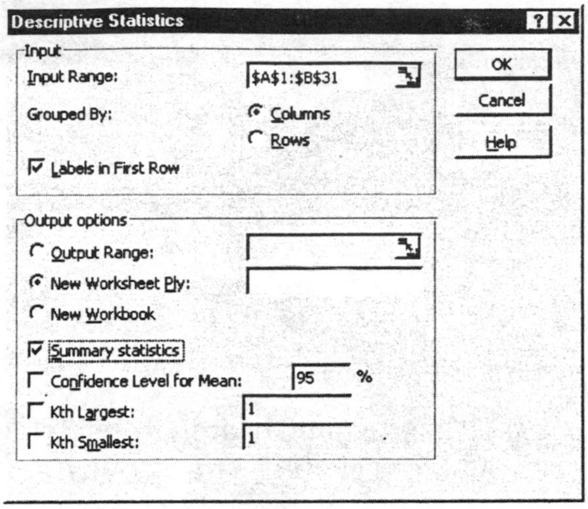

Descriptive Statistics

Input
Input Range: $A$1:$B$31
Grouped By: (•) Columns ( ) Rows
[x] Labels in First Row

Output options
( ) Output Range:
(•) New Worksheet Ply:
( ) New Workbook
[x] Summary statistics
[ ] Confidence Level for Mean: 95 %
[ ] Kth Largest: 1
[ ] Kth Smallest: 1

OK | Cancel | Help

4. You will want to make column A and column C wider so that the labels are easier to read. The variance of Time A is 0.2401 and the variance of Time B is 0.1075. Return to the worksheet that contains the data. To do this, click on the **Ex8_1-13** tab near the bottom of the screen.

	A	B	C	D
1	*TimeA*		*TimeB*	
2				
3	Mean	2.13	Mean	1.593333
4	Standard Error	0.089462	Standard Error	0.059872
5	Median	2.1	Median	1.6
6	Mode	2.1	Mode	1.6
7	Standard Deviation	0.490004	Standard Deviation	0.327933
8	Sample Variance	0.240103	Sample Variance	0.10754
9	Kurtosis	0.863836	Kurtosis	-0.57822
10	Skewness	-0.00957	Skewness	-0.25732
11	Range	2.3	Range	1.3
12	Minimum	1	Minimum	0.9
13	Maximum	3.3	Maximum	2.2
14	Sum	63.9	Sum	47.8
15	Count	30	Count	30

5. At the top of the screen, click **Tools** → **Data Analysis**. Select **z-Test: Two Sample for Means** and click **OK**.

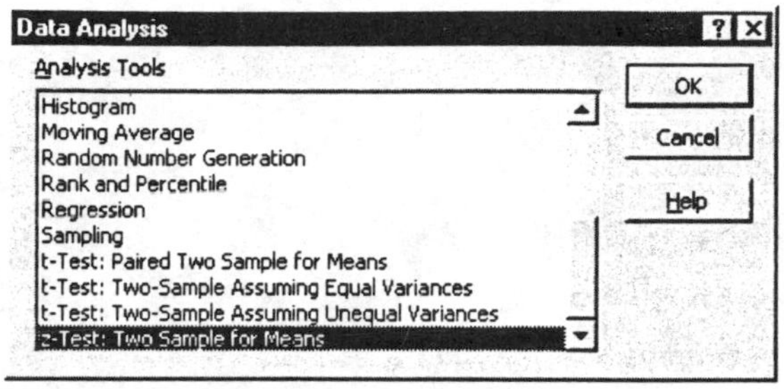

Data Analysis

Analysis Tools

Histogram
Moving Average
Random Number Generation
Rank and Percentile
Regression
Sampling
t-Test: Paired Two Sample for Means
t-Test: Two-Sample Assuming Equal Variances
t-Test: Two-Sample Assuming Unequal Variances
z-Test: Two Sample for Means

OK
Cancel
Help

6. Complete the z-Test: Two Sample for Means dialog box as shown below. Click **OK**.

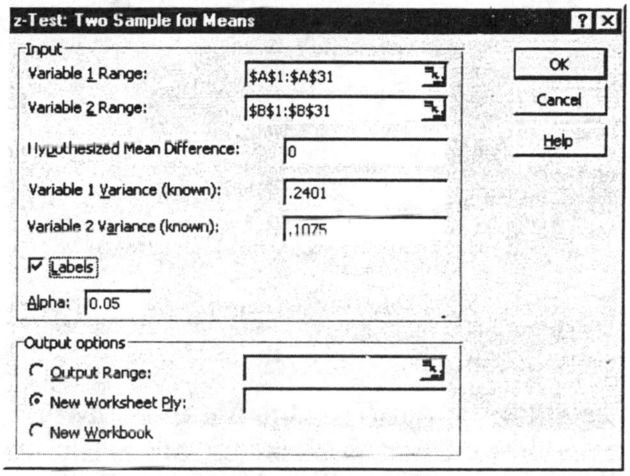

z-Test: Two Sample for Means

Input
Variable 1 Range: $A$1:$A$31
Variable 2 Range: $B$1:$B$31
Hypothesized Mean Difference: 0
Variable 1 Variance (known): .2401
Variable 2 Variance (known): .1075
Labels
Alpha: 0.05

Output options
Output Range:
New Worksheet Ply:
New Workbook

OK
Cancel
Help

The output is displayed in a new worksheet. You will want to adjust the width of column A so that you can read the labels. Your output should look similar to the output displayed below.

	A	B	C
1	z-Test: Two Sample for Means		
2			
3		*TimeA*	*TimeB*
4	Mean	2.13	1.593333
5	Known Variance	0.2401	0.1075
6	Observations	30	30
7	Hypothesized Mean Difference	0	
8	z	4.985691	
9	P(Z<=z) one-tail	3.09E-07	
10	z Critical one-tail	1.644853	
11	P(Z<=z) two-tail	6.18E-07	
12	z Critical two-tail	1.959961	

◀

Section 8.2

► Exercise 11 (pg. 385)

Testing the Claim That the Mean Footwell Intrusion for Small & Midsize Cars Is Equal

If the PHStat add-in has not been loaded, you will need to load it before continuing. Follow the instructions in Section GS 8.2.

1. First open a new Excel worksheet. Then, at the top of the screen, select **PHStat → Two-Sample Tests → t Test for Differences in Two Means.**

2. Complete the t Test for Differences in Two Means dialog box as shown below. Click **OK**.

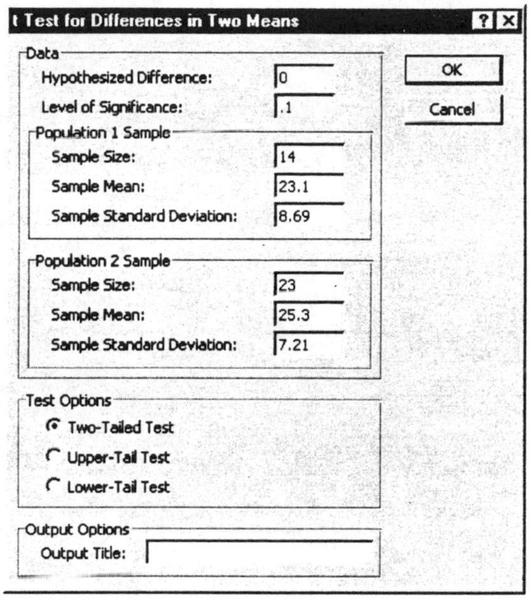

t Test for Differences in Two Means

Data

Hypothesized Difference: 0

Level of Significance: .1

Population 1 Sample

Sample Size: 14

Sample Mean: 23.1

Sample Standard Deviation: 8.69

Population 2 Sample

Sample Size: 23

Sample Mean: 25.3

Sample Standard Deviation: 7.21

Test Options

Two-Tailed Test

Upper-Tail Test

Lower-Tail Test

Output Options

Output Title:

OK

Cancel

The output is displayed in a new worksheet named "Hypothesis." You may need to scroll down a couple rows to see all of it.

	A	B
1	**t Test for Differences in Two Means**	
2		
3	**Hypothesized Difference**	**0**
4	**Level of Significance**	**0.1**
5	**Population 1 Sample**	
6	**Sample Mean**	**23.1**
7	**Sample Size**	**14**
8	**Sample Standard Deviation**	**8.69**
9	**Population 2 Sample**	
10	**Sample Mean**	**25.3**
11	**Sample Size**	**23**
12	**Sample Standard Deviation**	**7.21**
13	Population 1 Sample Degrees of Freedom	13
14	Population 2 Sample Degrees of Freedom	22
15	Total Degrees of Freedom	35
16	Pooled Variance	60.72456
17	Difference in Sample Means	-2.2
18	*t*-Test Statistic	-0.83285
19		
20	**Two-Tailed Test**	
21	**Lower Critical Value**	**-1.68957**
22	**Upper Critical Value**	**1.689573**

Hypothesis / Sheet1 / Sheet2 / Sheet3

◀

Section 8.3

▸ Example 2 (pg. 392) — The t-Test for the Difference Between Means

1. Open a new Excel worksheet and enter the golfers' scores as shown below.

	A	B
1	Old Design	New Design
2	89	83
3	84	83
4	96	92
5	82	84
6	74	76
7	92	91
8	85	80
9	91	91

2. At the top of the screen, click **Tools → Data Analysis**. Select **t-Test: Paired Two Sample for Means**. Click **OK**.

If Data Analysis does not appear as a choice in the Tools menu, you will need to load the Microsoft Excel ToolPak add-in. Follow the procedure in Section GS 8.1 before continuing

---

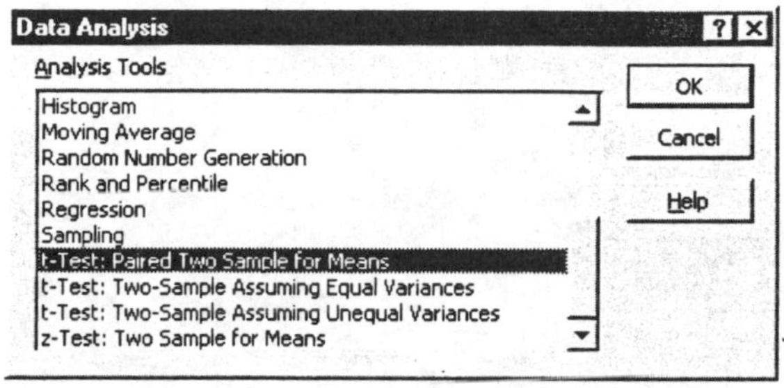

Data Analysis

Analysis Tools

Histogram
Moving Average
Random Number Generation
Rank and Percentile
Regression
Sampling
t-Test: Paired Two Sample for Means
t-Test: Two-Sample Assuming Equal Variances
t-Test: Two-Sample Assuming Unequal Variances
z-Test: Two Sample for Means

OK
Cancel
Help

3. Complete the t-Test: Paired Two Sample for Means dialog box as shown below. Click **OK**.

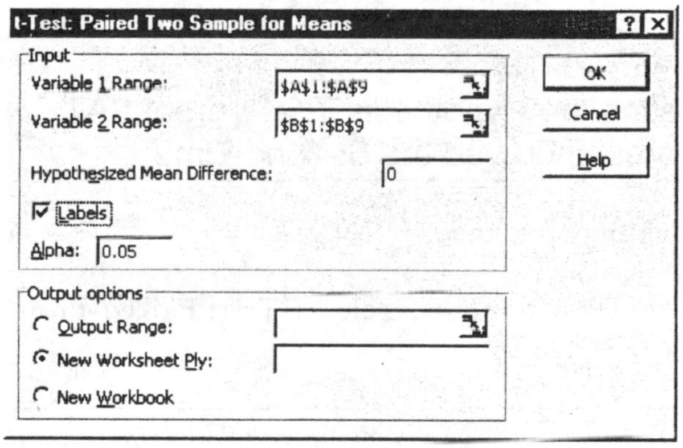

t-Test: Paired Two Sample for Means

Input
Variable 1 Range: $A$1:$A$9
Variable 2 Range: $B$1:$B$9
Hypothesized Mean Difference: 0
Labels
Alpha: 0.05

Output options
Output Range:
New Worksheet Ply:
New Workbook

OK
Cancel
Help

The output is displayed in a new worksheet. You will want to adjust the width of the columns so that you can read all the labels. Your output should look similar to the output displayed below.

	A	B	C
1	t-Test: Paired Two Sample for Means		
2			
3		*Old Design*	*New Design*
4	Mean	86.625	85
5	Variance	47.410714	33.714286
6	Observations	8	8
7	Pearson Correlation	0.896872	
8	Hypothesized Mean Difference	0	
9	df	7	
10	t Stat	1.4982596	
11	P(T<=t) one-tail	0.0888692	
12	t Critical one-tail	1.8945775	
13	P(T<=t) two-tail	0.1777385	
14	t Critical two-tail	2.3646226	

► Exercise 15 (pg. 396)

Testing the Hypothesis That Verbal SAT Scores Improved the Second Time

1. Open worksheet "Ex8_3-15" in the Chapter 8 folder.

2. At the top of the screen, click **Tools → Data Analysis**. Select **t-Test: Paired Two Sample for Means**. Click **OK**.

If Data Analysis does not appear as a choice in the Tools menu, you will need to load the Microsoft Excel ToolPak add-in. Follow the procedure in Section GS 8.1 before continuing.

Data Analysis

Analysis Tools

Histogram
Moving Average
Random Number Generation
Rank and Percentile
Regression
Sampling
t-Test: Paired Two Sample for Means
t-Test: Two-Sample Assuming Equal Variances
t-Test: Two-Sample Assuming Unequal Variances
z-Test: Two Sample for Means

OK
Cancel
Help

3. Complete the t-Test: Paired Two Sample for Means dialog box as shown below. Click **OK**.

t-Test: Paired Two Sample for Means

Input
Variable 1 Range: $A$1:$A$15
Variable 2 Range: $B$1:$B$15
Hypothesized Mean Difference: 0
☑ Labels
Alpha: 0.01

Output options
○ Output Range:
◉ New Worksheet Ply:
○ New Workbook

OK
Cancel
Help

The output is displayed in a new worksheet. You will want to make the columns wider so that you can read all the labels. Your output should appear similar to the output shown below.

	A	B	C
1	t-Test: Paired Two Sample for Means		
2			
3		*First SAT*	*Second SAT*
4	Mean	483.6429	517.357143
5	Variance	8372.555	5060.09341
6	Observations	14	14
7	Pearson Correlation	0.896142	
8	Hypothesized Mean Difference	0	
9	df	13	
10	t Stat	-3.0011	
11	P(T<=t) one-tail	0.005109	
12	t Critical one-tail	2.650304	
13	P(T<=t) two-tail	0.010217	
14	t Critical two-tail	3.012283	

► Exercise 23 (pg. 399)

Constructing a 90% Confidence Interval for μ_D, the Mean Increase in Hours of Sleep

1. Open worksheet "EX8_3-23" in the Chapter 8 folder.

2. Use Excel to calculate a difference score for each of the 12 patients. First, click in cell D1 and key in the label **Difference.**

	A	B	C	D
1	Without dr	With new drug		Difference
2	1.8	3		

3. Click in cell D2 and enter the formula **=A2-B2** as shown below. Press [**Enter**].

	A	B	C	D	E
1	Without dr	With new drug		Difference	
2	1.8	3		=A2-B2	

4. Click in cell **D2** (where –1.2 now appears) and copy the contents of that cell to cells D3 through D13.

	A	B	C	D
1	Without dr	With new drug		Difference
2	1.8	3		-1.2
3	2	3.6		-1.6
4	3.4	4		-0.6
5	3.5	4.4		-0.9
6	3.7	4.5		-0.8
7	3.8	5.2		-1.4
8	3.9	5.5		-1.6
9	3.9	5.7		-1.8
10	4	6.2		-2.2
11	4.9	6.3		-1.4
12	5.1	6.6		-1.5
13	5.2	7.8		-2.6

5. You will now obtain descriptive statistics for Difference. At the top of the screen, click **Tools → Data Analysis**. Select **Descriptive Statistics**. Click **OK**.

If Data Analysis does not appear as a choice in the Tools menu, you will need to load the Microsoft Excel ToolPak add-in. Follow the procedure in Section GS 8.1 before continuing.

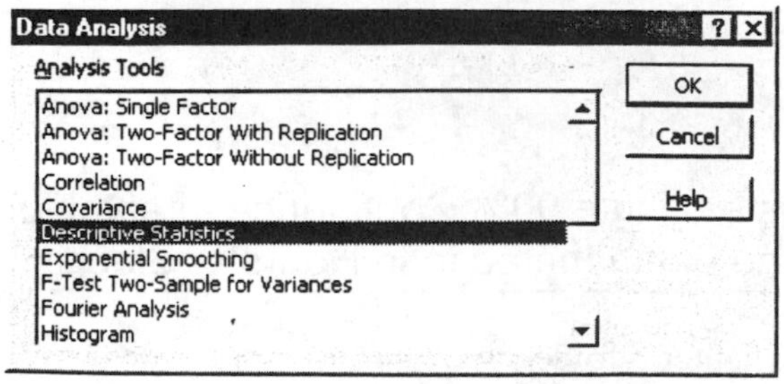

Data Analysis

Analysis Tools

Anova: Single Factor
Anova: Two-Factor With Replication
Anova: Two-Factor Without Replication
Correlation
Covariance
Descriptive Statistics
Exponential Smoothing
F-Test Two-Sample for Variances
Fourier Analysis
Histogram

OK
Cancel
Help

6. Complete the Descriptive Statistics dialog box as shown below. Click **OK**.

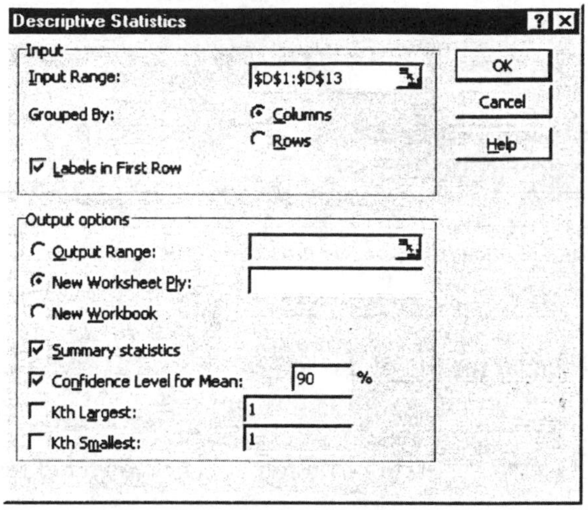

Descriptive Statistics

Input
- Input Range: $D$1:$D$13
- Grouped By: (•) Columns ( ) Rows
- [x] Labels in First Row

Output options
- ( ) Output Range:
- (•) New Worksheet Ply:
- ( ) New Workbook
- [x] Summary statistics
- [x] Confidence Level for Mean: 90 %
- [ ] Kth Largest: 1
- [ ] Kth Smallest: 1

OK | Cancel | Help

The output is displayed in a new worksheet. To calculate the lower limit of the 90% confidence interval, subtract 0.295204 from –1.46667. To calculate the upper limit, add 0.295204 to –1.46667.

	A	B
1	*Difference*	
2		
3	Mean	-1.46667
4	Standard Error	0.164378
5	Median	-1.45
6	Mode	-1.6
7	Standard Deviation	0.569423
8	Sample Variance	0.324242
9	Kurtosis	0.196173
10	Skewness	-0.43907
11	Range	2
12	Minimum	-2.6
13	Maximum	-0.6
14	Sum	-17.6
15	Count	12
16	Confidence Level(90.0%)	0.295204

Section 8.4

► Example 1 (pg. 402) A Two-Sample z-Test for the Difference Between Proportions

If the PHStat add-in has not been loaded, you will need to load it before continuing. Follow the instructions in Section GS 8.2.

1. First open a new Excel worksheet. Then, at the top of the screen, select **PHStat → Two-Sample Tests → Z Test for Differences in Two Proportions**.

2. Complete the Z Test for Differences in Two Proportions dialog box as shown below. Click **OK**.

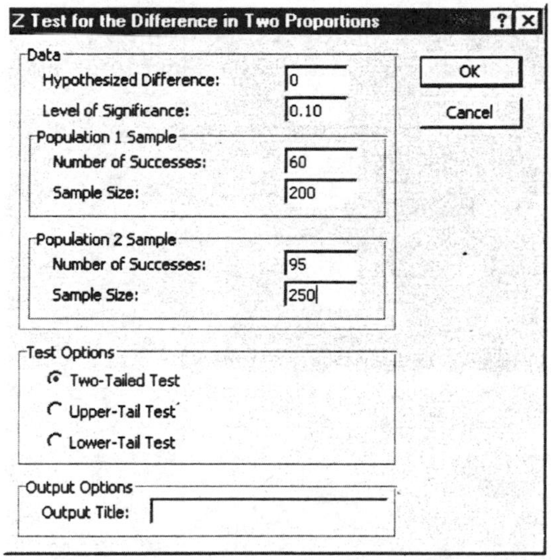

Z Test for the Difference in Two Proportions

Data
Hypothesized Difference: 0
Level of Significance: 0.10
Population 1 Sample
Number of Successes: 60
Sample Size: 200
Population 2 Sample
Number of Successes: 95
Sample Size: 250

Test Options
Two-Tailed Test
Upper-Tail Test
Lower-Tail Test

Output Options
Output Title:

OK
Cancel

The output is displayed in a worksheet named "Hypothesis."

	A	B
1	**Z Test for Differences in Two Proportions**	
2		
3	**Hypothesized Difference**	**0**
4	**Level of Significance**	**0.1**
5	**Group 1**	
6	**Number of Successes**	**60**
7	**Sample Size**	**200**
8	**Group 2**	
9	**Number of Successes**	**95**
10	**Sample Size**	**250**
11	Group 1 Proportion	0.3
12	Group 2 Proportion	0.38
13	Difference in Two Proportions	-0.08
14	Average Proportion	0.344444444
15	Z Test Statistic	-1.774615984
16		
17	**Two-Tailed Test**	
18	**Lower Critical Value**	**-1.644853**
19	**Upper Critical Value**	**1.644853**
20	***p*-Value**	**0.075961219**
21	**Reject the null hypothesis**	

◀

► Exercise 7 (pg. 404) Testing the Claim That the Use of Alternative Medicines Has Not Changed

If the PHStat add-in has not been loaded, you will need to load it before continuing. Follow the instructions in Section GS 8.2.

1. First open a new Excel worksheet. Then, at the top of the screen, select **PHStat → Two-Sample Tests → Z Test for Differences in Two Proportions**.

2. Complete the Z Test for Differences in Two Proportions dialog box as shown below. Click **OK**.

Z Test for the Difference in Two Proportions

Data
- Hypothesized Difference: 0
- Level of Significance: 0.05

Population 1 Sample
- Number of Successes: 520
- Sample Size: 1539

Population 2 Sample
- Number of Successes: 865
- Sample Size: 2055

Test Options
- (•) Two-Tailed Test
- ( ) Upper-Tail Test
- ( ) Lower-Tail Test

Output Options
- Output Title:

OK | Cancel

The output is displayed in a worksheet named "Hypothesis."

	A	B
1	**Z Test for Differences in Two Proportions**	
2		
3	**Hypothesized Difference**	**0**
4	**Level of Significance**	**0.05**
5	**Group 1**	
6	**Number of Successes**	**520**
7	**Sample Size**	**1539**
8	**Group 2**	
9	**Number of Successes**	**865**
10	**Sample Size**	**2055**
11	Group 1 Proportion	0.337881741
12	Group 2 Proportion	0.420924574
13	Difference in Two Proportions	-0.083042833
14	Average Proportion	0.385364496
15	Z Test Statistic	-5.06166817
16		
17	**Two-Tailed Test**	
18	**Lower Critical Value**	**-1.959961082**
19	**Upper Critical Value**	**1.959961082**
20	***p*-Value**	**4.16302E-07**
21	**Reject the null hypothesis**	

◄

Technology Lab

► Exercise 1 (pg. 407)

Testing the Hypothesis That the Probability of a "Found Coin" Lying Heads Up is 0.5

If the PHStat add-in has not been loaded, you will need to load it before continuing. Follow the instructions in Section GS 8.2.

1. First open a new Excel worksheet. Then, at the top of the screen, select **PHStat → One-Sample Tests → Z Test for the Proportion.**

2. Complete the Z Test for the Proportion dialog box as shown below. Click **OK**.

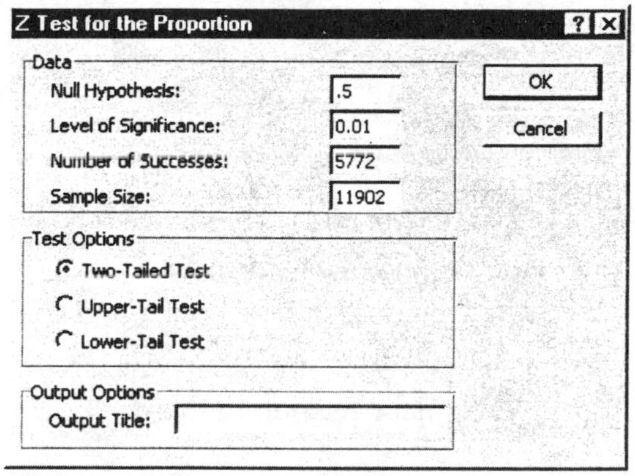

Z Test for the Proportion

Data
Null Hypothesis: .5
Level of Significance: 0.01
Number of Successes: 5772
Sample Size: 11902

OK
Cancel

Test Options
Two-Tailed Test
Upper-Tail Test
Lower-Tail Test

Output Options
Output Title:

Your output should look similar to the output displayed below.

	A	B	C
1	**Z Test of Hypothesis for the Proportion**		
2			
3	**Null Hypothesis** *p*=	**0.5**	
4	**Level of Significance**	**0.01**	
5	**Number of Successes**	**5772**	
6	**Sample Size**	**11902**	
7	Sample Proportion	0.484960511	
8	Standard Error	0.004583107	
9	Z Test Statistic	-3.281504874	
10			
11	**Two-Tailed Test**		
12	**Lower Critical Value**	**-2.575834515**	
13	**Upper Critical value**	**2.575834515**	
14	***p*-Value**	**0.001032667**	
15	**Reject the null hypothesis**		

◀

► Exercise 3 (pg. 407) Simulating "Tails Over Heads"

1. Open a new Excel worksheet and click in cell **A1** to place the output there.

2. At the top of the screen, select **Tools** → **Data Analysis**. Select **Random Number Generation**. Click **OK**.

3. Complete the Random Number Generation dialog box as shown below. Click **OK**.

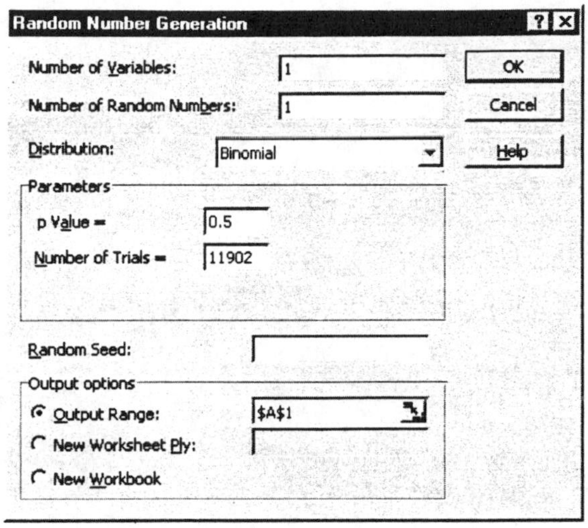

Random Number Generation

Number of Variables: 1

Number of Random Numbers: 1

Distribution: Binomial

Parameters

p Value = 0.5

Number of Trials = 11902

Random Seed:

Output options

Output Range: $A$1

New Worksheet Ply:

New Workbook

OK

Cancel

Help

The output for this simulation indicates that 5,955 of the 11,902 coins were found heads up. Because this output was generated randomly, it is unlikely that your output will be exactly the same.

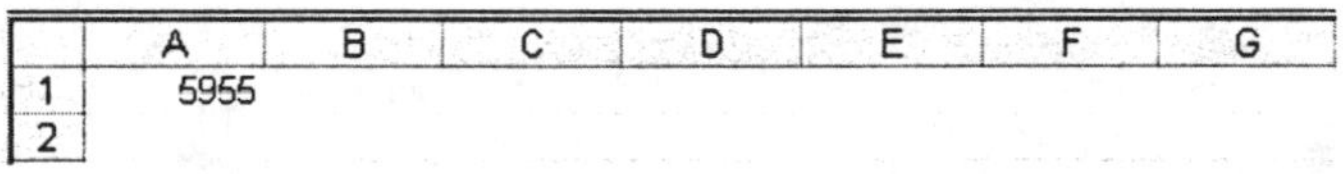

	A	B	C	D	E	F	G
1	5955						
2							

◀

Correlation and Regression

Section 9.1

► Example 3 (pg. 420) Constructing a Scatter Plot

1. Open worksheet "OldFaithful" in the Chapter 9 folder.

2. Click on any cell within the table of data. At the top of the screen, click **Insert** and select **Chart** from the menu that appears.

3. In the Chart Type dialog box, select the **XY (Scatter)** chart type by clicking on it. Then click on the topmost chart sub-type. Click **Next**.

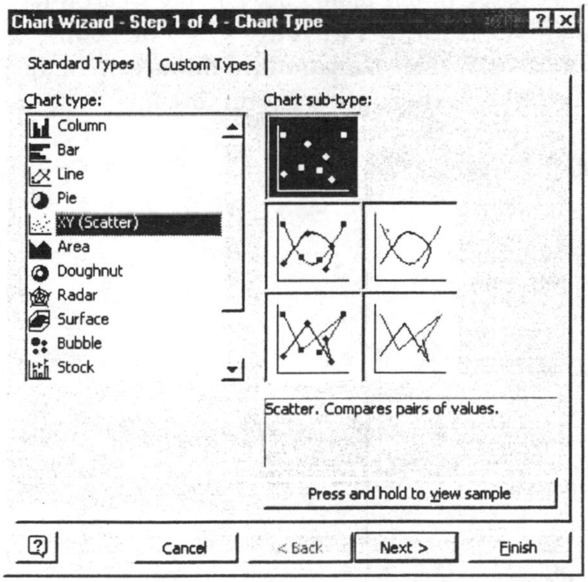

4. In the Chart Source Data dialog box, the entry in the data range field should be **=OldFaithful!$A$1:$B$36**. Make any necessary corrections. Then click **Next**.

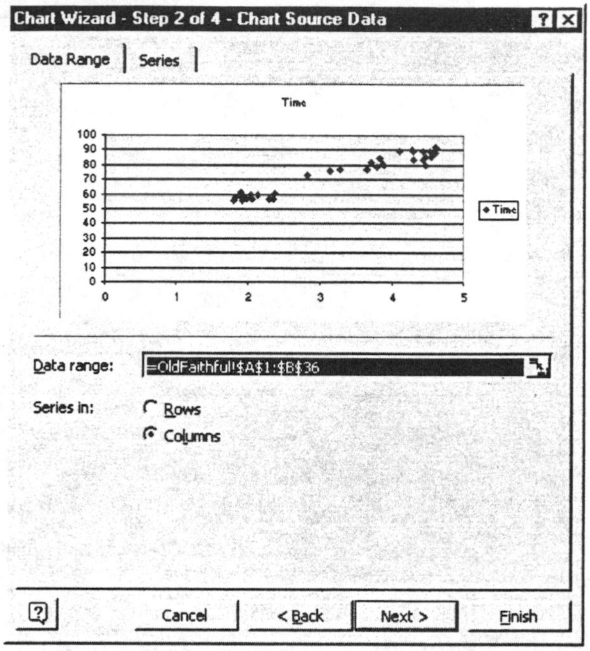

5. Click the **Titles** tab at the top of the Chart Options dialog box. In the Chart title field, change the title to **Duration of Old Faithful's Eruptions by Time Until the Next Eruption**. In the Value (X) axis field, enter **Duration (in minutes)**. In the Value (Y) axis field, enter **Time until the next eruption (in minutes)**.

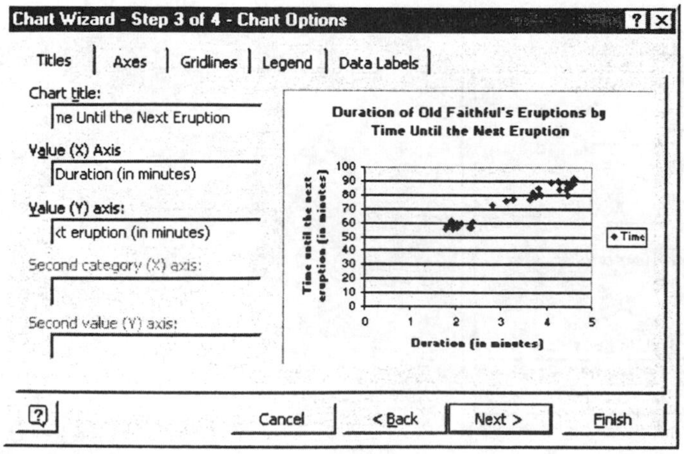

6. Click the **Legend** tab at the top of the Chart Options dialog box. Click in the box to the left of **Show Legend** so that a checkmark does not appear there. This removes the Time legend from the right side of the scatter plot. Click **Next.**

Chart Wizard - Step 3 of 4 - Chart Options
Titles | Axes | Gridlines | Legend | Data Labels
Show legend
Placement
Bottom
Corner
Top
Right
Left
Duration of Old Faithful's Eruptions by Time Until the Next Eruption
Time until the next eruption (in minutes)
100 90 80 70 60 50 40 30 20 10 0
0 1 2 3 4 5
Duration (in minutes)
Cancel | < Back | Next > | Finish

7. The Chart Location dialog box presents two options for placement of the scatter plot. For this example, select **As object in** the same worksheet as the data. To do this, click the button to the left of **As object in** so that a black dot appears there. Click **Finish**.

Chart Wizard - Step 4 of 4 - Chart Location
Place chart:
As new sheet: Chart1
As object in: OldFaithful
Cancel | < Back | Next > | Finish

8. Make the chart taller so that it is easier to read. To do this, first click within the figure near a border. Black square handles appear. Then click on the center handle on the bottom border of the figure and drag it down a few rows.

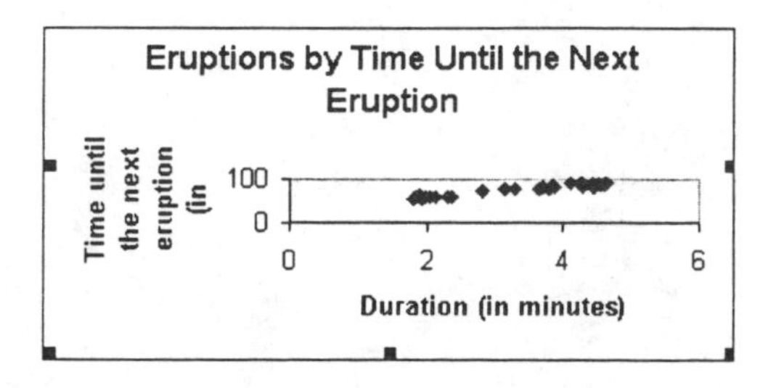

Your scatter plot should now look similar to the one shown below.

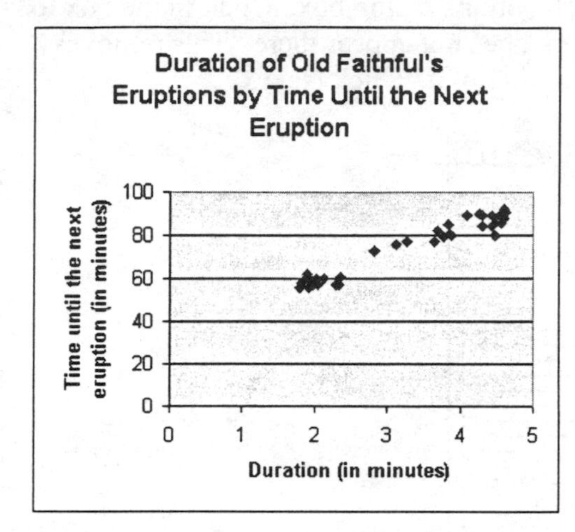

◀

► Example 5 (pg. 423)

Finding a Correlation Coefficient

1. Open worksheet "OldFaithful" in the Chapter 9 folder and click in cell C37 to place the output there.

	A	B	C	D	E	F	G
35	4.6	92					
36	4.63	91					
37							

2. At the top of the screen, click **Insert → Function.**

3. Under Function category, select **Statistical**. Under Function name, select **CORREL**. Click **OK**.

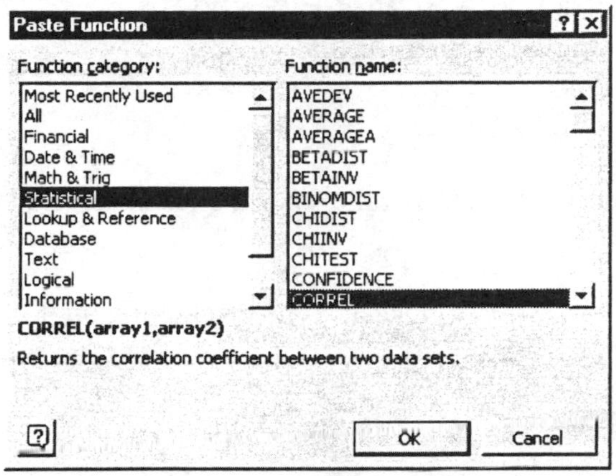

Paste Function

Function category:
Most Recently Used
All
Financial
Date & Time
Math & Trig
Statistical
Lookup & Reference
Database
Text
Logical
Information

Function name:
AVEDEV
AVERAGE
AVERAGEA
BETADIST
BETAINV
BINOMDIST
CHIDIST
CHIINV
CHITEST
CONFIDENCE
CORREL

CORREL(array1,array2)

Returns the correlation coefficient between two data sets.

OK Cancel

4. Complete the CORREL dialog box as shown below. Click **OK**.

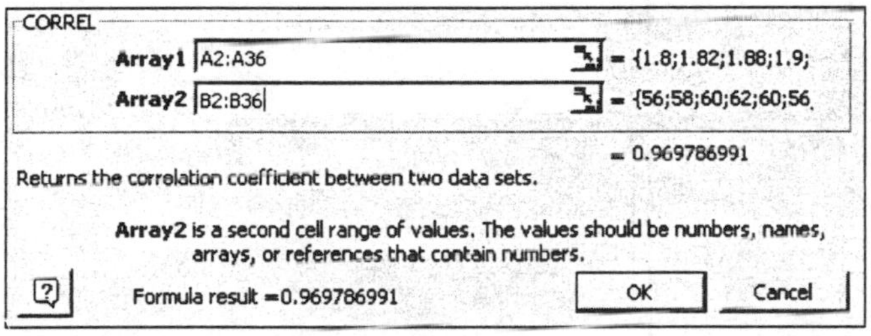

CORREL

Array1 A2:A36 = {1.8;1.82;1.88;1.9;

Array2 B2:B36 = {56;58;60;62;60;56

= 0.969786991

Returns the correlation coefficient between two data sets.

Array2 is a second cell range of values. The values should be numbers, names, arrays, or references that contain numbers.

Formula result =0.969786991 OK Cancel

Your output should look like the output shown below.

	A	B	C	D	E	F	G
35	4.6	92					
36	4.63	91					
37			0.969787				

◀

► Exercise 13 (pg. 428) Constructing a Scatter Plot and Determining Correlation

1. Open worksheet "Ex9_1-13" in the Chapter 9 folder.

2. Click on any cell within the table of data. At the top of the screen, click **Insert** and select **Chart** from the menu that appears.

3. In the Chart Type dialog box, select the **XY (Scatter)** chart type by clicking on it. Then click on the topmost chart sub-type. Click **Next**.

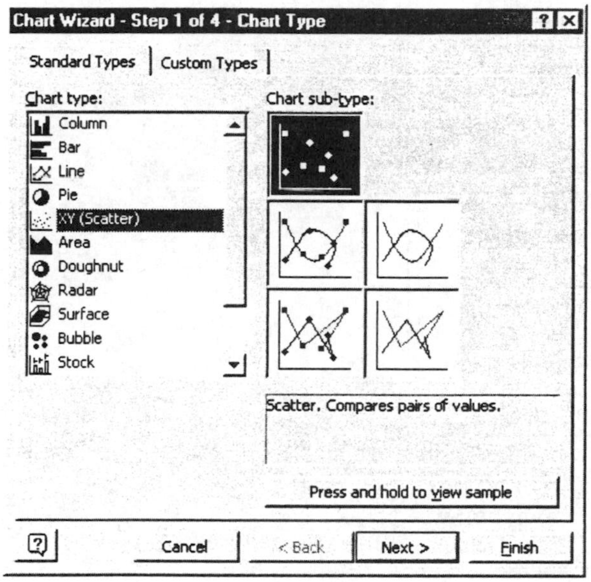

4. In the Chart Source Data dialog box, the entry in the data range field should be **='Ex9_1-13'!$A$1:$B$14**. Make any necessary corrections. Then click **Next**.

5. Click the **Titles** tab at the top of the Chart Options dialog box. In the Chart title field, change the title to **Hours Spent Studying for a Test by Test Score**. In the Value (X) axis field, enter **Hours**. In the Value (Y) axis field, enter **Test Score**.

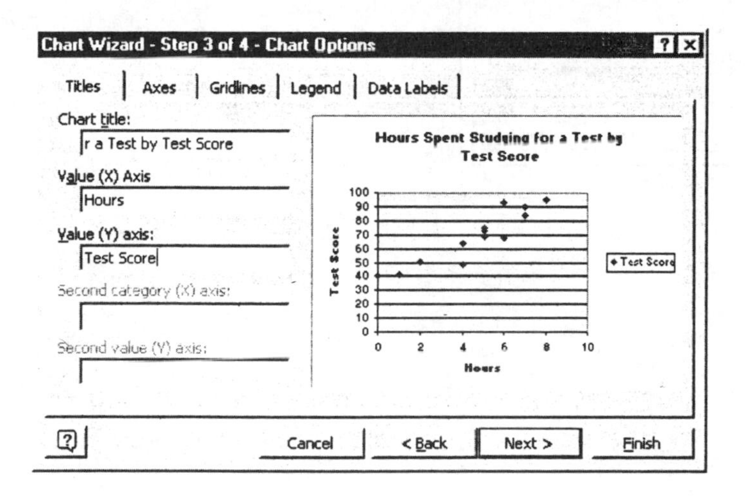

6. Click the **Legend** tab at the top of the Chart Options dialog box. Click in the box to the left of **Show Legend** so that a checkmark does not appear there. This removes the Test Score legend from the right side of the scatter plot. Click **Next**.

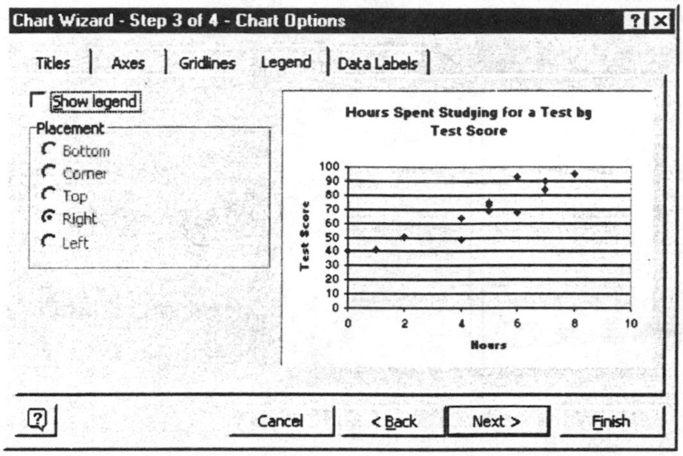

7. The Chart Location dialog box presents two options for placement of the scatter plot. For this example, select **As object in** the same worksheet as the data. To do this, click the button to the left of **As object in** so that a black dot appears there. Click **Finish**.

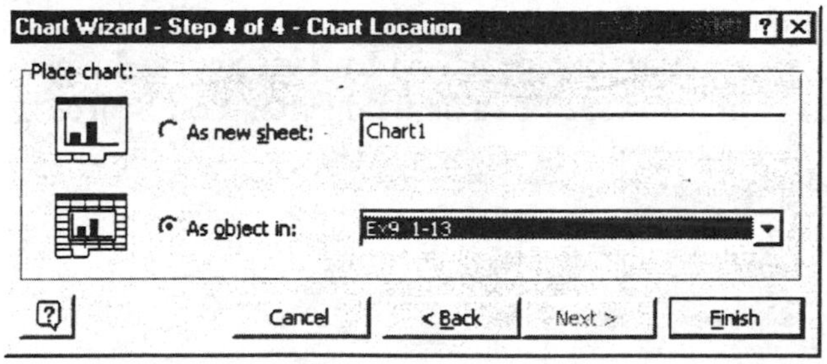

8. Make the chart taller so that it is easier to read. To do this, first click within the figure near a border. Black square handles appear. Then click on the center handle on the bottom border of the figure and drag it down a few rows.

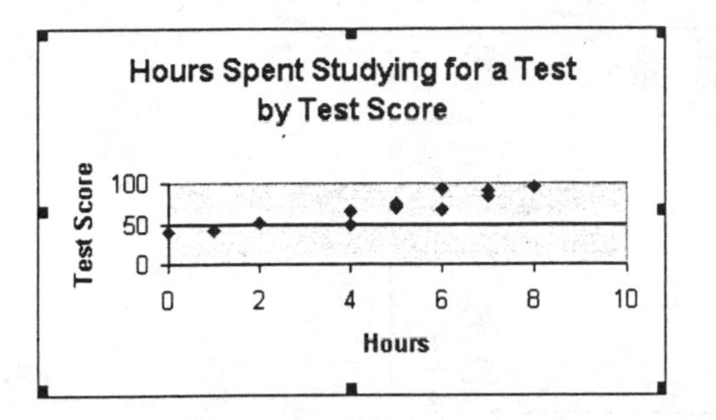

Your scatter plot should now look similar to the one shown below.

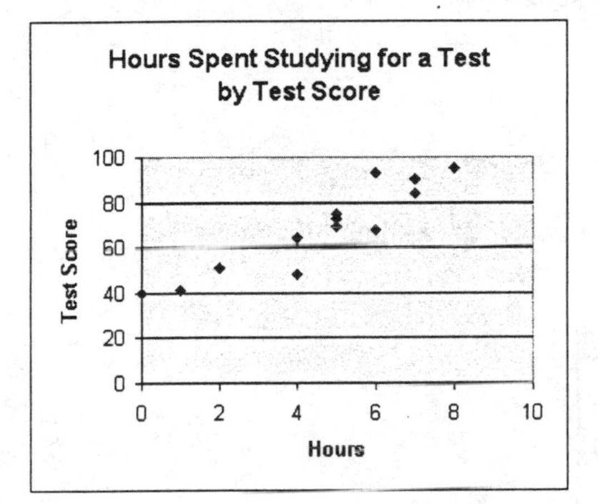

9. You will now use Excel to find the correlation between hours and test score. Click in cell **B16** of the worksheet to place the output there.

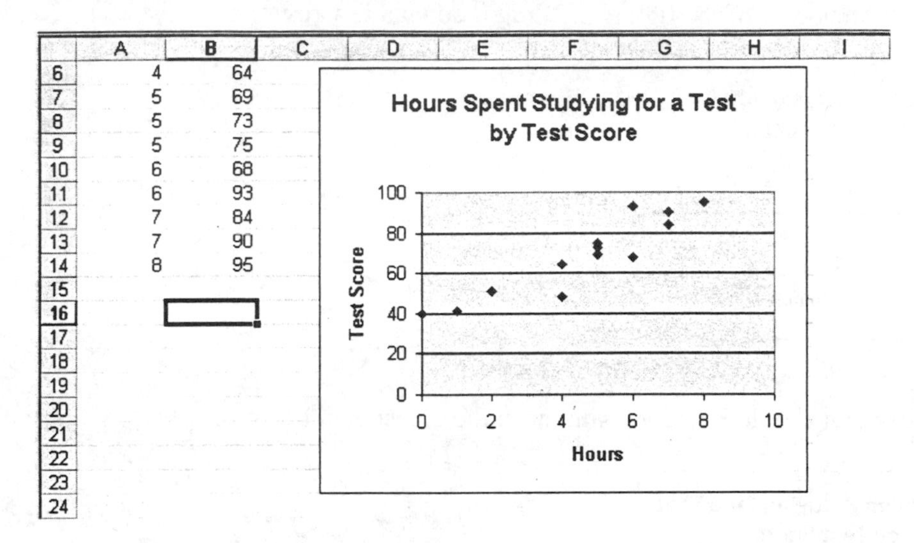

10. At the top of the screen, click **Insert → Function**.

11. Under Function category, select **Statistical**. Under Function name, select **CORREL**. Click **OK**.

Paste Function

Function category:
Most Recently Used
All
Financial
Date & Time
Math & Trig
Statistical
Lookup & Reference
Database
Text
Logical
Information

Function name:
AVEDEV
AVERAGE
AVERAGEA
BETADIST
BETAINV
BINOMDIST
CHIDIST
CHIINV
CHITEST
CONFIDENCE
CORREL

CORREL(array1,array2)

Returns the correlation coefficient between two data sets.

OK Cancel

12. Complete the CORREL dialog box as shown below. Click **OK**.

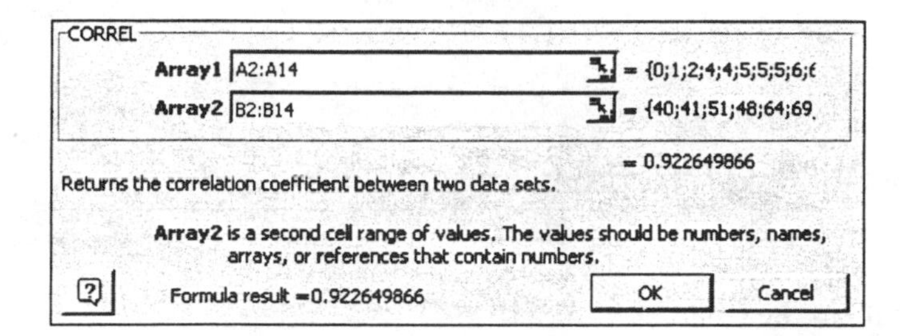

Your output should look like the output shown below.

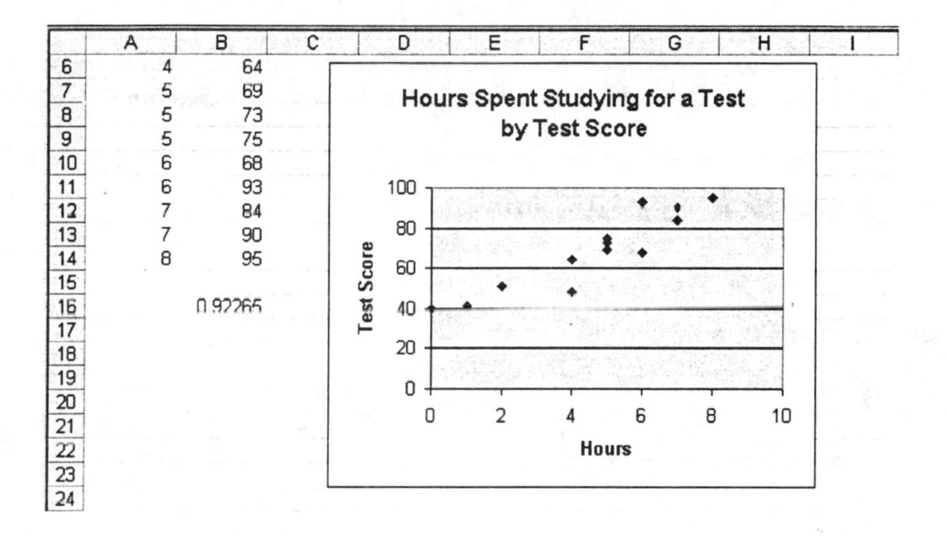

	A	B
6	4	64
7	5	69
8	5	73
9	5	75
10	6	68
11	6	93
12	7	84
13	7	90
14	8	95
15		
16		0.92265

Section 9.2

▶ Example 2 (pg. 434) Finding a Regression Equation

1. Open worksheet "OldFaithful" in the Chapter 9 folder.

2. At the top of the screen, click **Tools → Data Analysis**. Select **Regression** and click **OK**.

If Data Analysis does not appear as a choice in the Tools menu, you will need to load the Microsoft Excel Analysis ToolPak add-in. Follow the procedure in Section GS 8.1 before continuing.

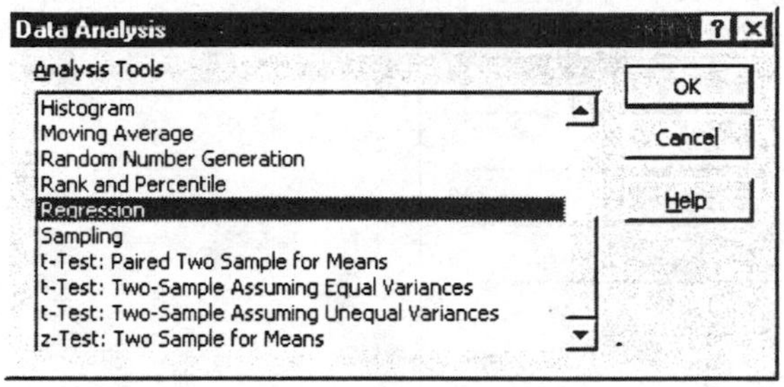

Data Analysis

Analysis Tools

Histogram
Moving Average
Random Number Generation
Rank and Percentile
Regression
Sampling
t-Test: Paired Two Sample for Means
t-Test: Two-Sample Assuming Equal Variances
t-Test: Two-Sample Assuming Unequal Variances
z-Test: Two Sample for Means

OK
Cancel
Help

3. Complete the Regression dialog box as shown below. Click **OK**.

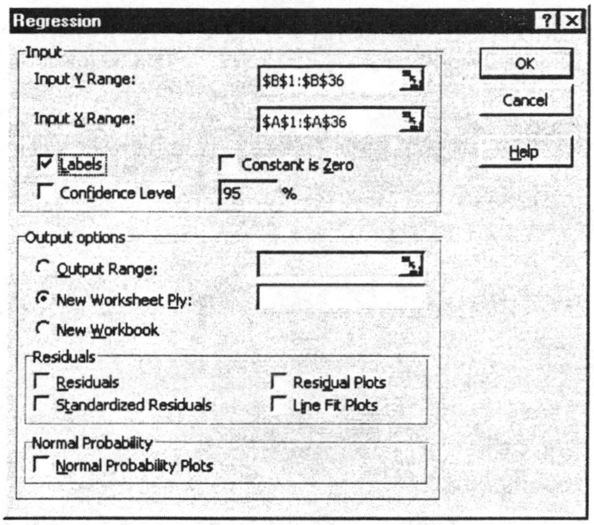

Regression

Input
Input Y Range: $B$1:$B$36
Input X Range: $A$1:$A$36
☑ Labels ☐ Constant is Zero
☐ Confidence Level 95 %

Output options
○ Output Range:
◉ New Worksheet Ply:
○ New Workbook

Residuals
☐ Residuals ☐ Residual Plots
☐ Standardized Residuals ☐ Line Fit Plots

Normal Probability
☐ Normal Probability Plots

OK
Cancel
Help

You will want to make some of the columns wider so that you can read all the labels. Your output should look similar to the output shown below. The intercept and slope of the regression equation are shown in the bottom two lines of the output under the label "Coefficients." The intercept is 35.301 and the slope is 11.824.

	A	B	C	D	E	F	G	H	I
16		*Coefficients*	*Standard Error*	*t Stat*	*P-value*	*Lower 95%*	*Upper 95%*	*Lower 95.0%*	*Upper 95.0%*
17	Intercept	35.301171	1.804267929	19.56537	1E-19	31.630357	38.97198	31.630357	38.9719847
18	Duration	11.824408	0.517788852	22.83635	8.57E-22	10.770958	12.87786	10.770958	12.877858

◀

► Exercise 11 (pg. 437)

Finding the Equation of a Regression Line

1. Open a new Excel worksheet and enter the age and systolic blood pressure data as shown below.

	A	B	C	D	E	F	G
1	Age	Blood Pressure					
2	16	109					
3	25	122					
4	39	143					
5	45	132					
6	49	199					
7	64	185					
8	70	199					

2. At the top of the screen, click **Tools → Data Analysis**. Select **Regression** and click **OK**.

If Data Analysis does not appear as a choice in the Tools menu, you will need to load the Microsoft Excel Analysis ToolPak add-in. Follow the procedure in Section GS 8.1 before continuing.

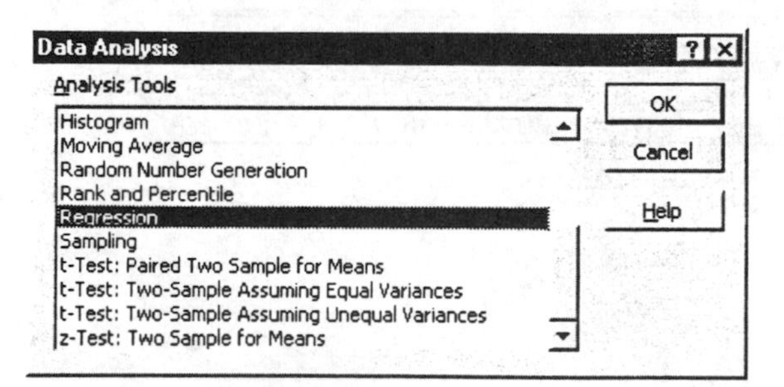

3. Complete the Regression dialog box as shown below. Click **OK**.

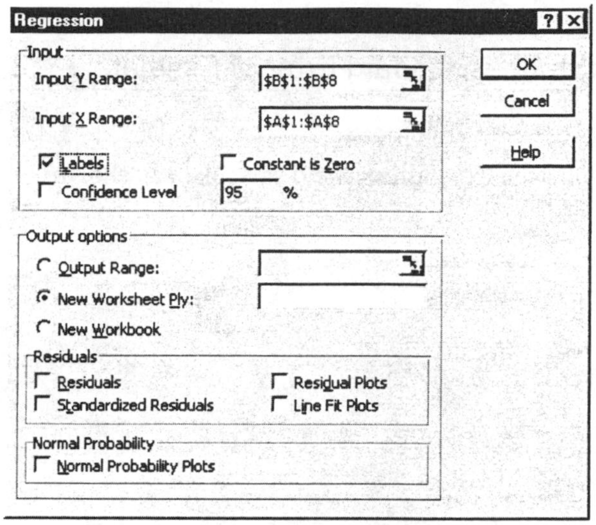

Regression

Input
Input Y Range: $B$1:$B$8
Input X Range: $A$1:$A$8
☑ Labels ☐ Constant is Zero
☐ Confidence Level 95 %

OK
Cancel
Help

Output options
○ Output Range:
◉ New Worksheet Ply:
○ New Workbook

Residuals
☐ Residuals ☐ Residual Plots
☐ Standardized Residuals ☐ Line Fit Plots

Normal Probability
☐ Normal Probability Plots

You will want to make some of the columns wider so that you can read all the labels. Your output should look similar to the output shown below. The intercept and slope of the regression equation are shown in the bottom two lines of the output under the label "Coefficients." The intercept is 79.733 and the slope is 1.724.

	A	B	C	D	E	F	G	H	I
16		*Coefficients*	*Standard Error*	*t Stat*	*P-value*	*Lower 95%*	*Upper 95%*	*Lower 95.0%*	*Upper 95.0%*
17	Intercept	79.7334	19.43492651	4.102583	0.009331	29.77441	129.6924	29.774413	129.692388
18	Age	1.7235915	0.408765078	4.216582	0.008355	0.672829	2.774354	0.67282918	2.77435392

◀

Section 9.3

► Example 2 (pg. 445) Finding the Standard Error of Estimate

1. Open a new Excel worksheet and enter the expenditures and sales data as shown below.

	A	B	C	D	E	F	G
1	Expenditures	Sales					
2	2.4	225					
3	1.6	184					
4	2	220					
5	2.6	240					
6	1.4	180					
7	1.6	184					
8	2	186					
9	2.2	215					

2. At the top of the screen, click **Tools → Data Analysis**. Select **Regression** and click **OK.**

If Data Analysis does not appear as a choice in the Tools menu, you will need to load the Microsoft Excel Analysis ToolPak add-in. Follow the procedure in Section GS 8.1 before continuing.

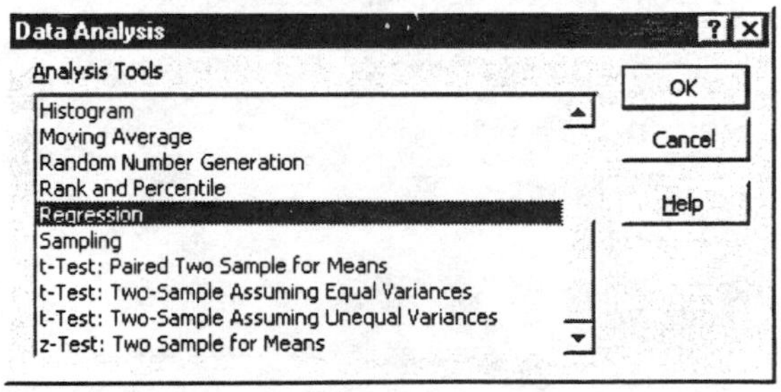

Data Analysis ? X
Analysis Tools
Histogram
Moving Average
Random Number Generation
Rank and Percentile
Regression
Sampling
t-Test: Paired Two Sample for Means
t-Test: Two-Sample Assuming Equal Variances
t-Test: Two-Sample Assuming Unequal Variances
z-Test: Two Sample for Means
OK
Cancel
Help

3. Complete the Regression dialog box as shown below. Click **OK**.

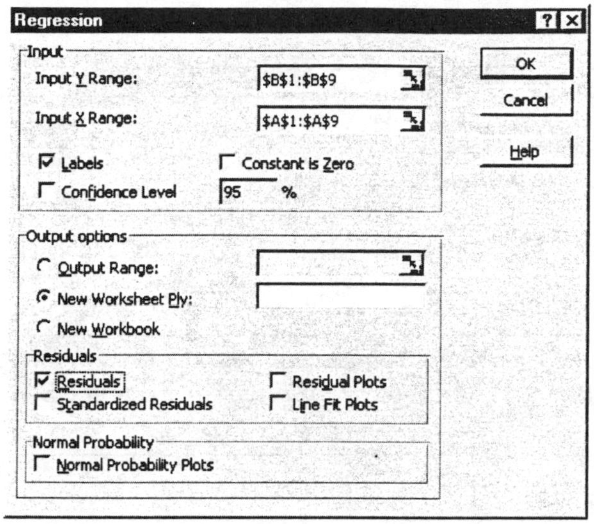

Regression
Input
Input Y Range: $B$1:$B$9
Input X Range: $A$1:$A$9
Labels
Constant is Zero
Confidence Level 95 %
OK
Cancel
Help
Output options
Output Range:
New Worksheet Ply:
New Workbook
Residuals
Residuals
Residual Plots
Standardized Residuals
Line Fit Plots
Normal Probability
Normal Probability Plots

The standard error of estimate is displayed in the upper portion of the output. You will want to widen column A so that you can read all the labels. The standard error of estimate is equal to 10.2903.

	A	B
1	SUMMARY OUTPUT	
2		
3	*Regression Statistics*	
4	Multiple R	0.91290528
5	R Square	0.83339606
6	Adjusted R Square	0.80562874
7	Standard Error	10.2903201
8	Observations	8

◀

Section 9.4

▶ Example 1 (pg. 452) Finding a Multiple Regression Equation

1. Open worksheet "Salary" in the Chapter 9 folder.

2. At the top of the screen, click **Tools → Data Analysis**. Select **Regression** and click **OK**.

If Data Analysis does not appear as a choice in the Tools menu, you will need to load the Microsoft Excel Analysis ToolPak add-in. Follow the procedure in Section GS 8.1 before continuing.

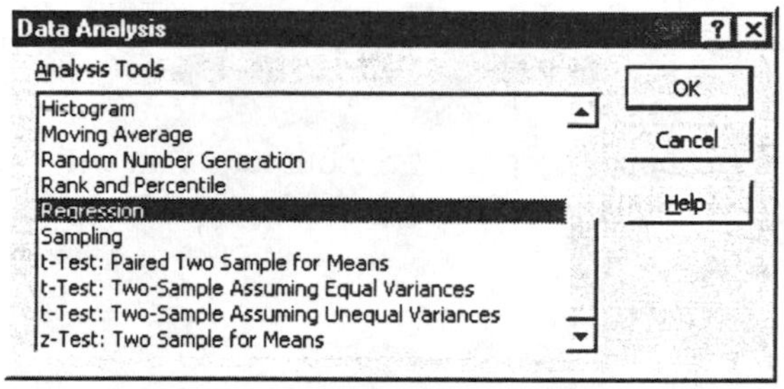

Data Analysis
Analysis Tools
Histogram
Moving Average
Random Number Generation
Rank and Percentile
Regression
Sampling
t-Test: Paired Two Sample for Means
t-Test: Two-Sample Assuming Equal Variances
t-Test: Two-Sample Assuming Unequal Variances
z-Test: Two Sample for Means
OK
Cancel
Help

3. Complete the Regression dialog box as shown below. Click **OK**.

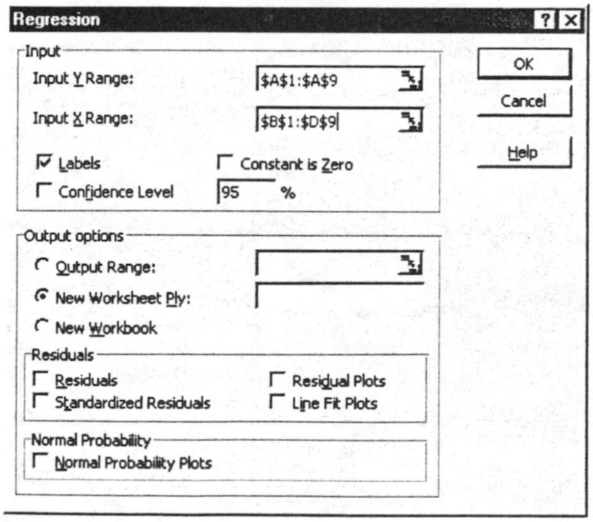

Regression
Input
Input Y Range: $A$1:$A$9
Input X Range: $B$1:$D$9
Labels
Constant is Zero
Confidence Level 95 %
Output options
Output Range:
New Worksheet Ply:
New Workbook
Residuals
Residuals
Standardized Residuals
Residual Plots
Line Fit Plots
Normal Probability
Normal Probability Plots
OK
Cancel
Help

The coefficients for the multiple regression equation are displayed in the lower portion of the output under the label "Coefficients."

	A	B	C	D	E	F	G	H	I
16		*Coefficients*	*Standard Error*	*t Stat*	*P-value*	*Lower 95%*	*Upper 95%*	*Lower 95.0%*	*Upper 95.0%*
17	Intercept	29764.446	1981.346465	15.02233	0.000114	24263.33	35265.56	24263.3349	35265.5571
18	Employment (yrs)	364.41203	48.31750816	7.542029	0.001656	230.2608	498.5632	230.260841	498.563215
19	Experience (yrs)	227.61881	123.8361513	1.838064	0.139912	-116.206	571.4438	-116.20618	571.443799
20	Education (yrs)	266.93504	147.3556227	1.811502	0.144295	-142.191	676.0607	-142.1906	676.060686

◀

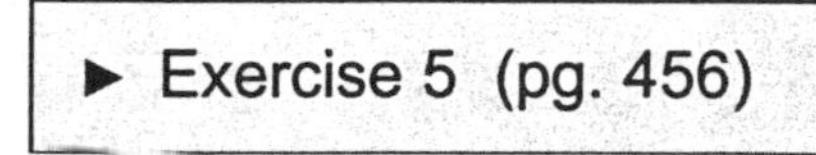

Finding a Multiple Regression Equation, the Standard Error of Estimate, and R^2

1. Open worksheet "ex9_4-5" in the Chapter 9 folder.

2. At the top of the screen, click **Tools → Data Analysis**. Select **Regression** and click **OK**.

If Data Analysis does not appear as a choice in the Tools menu, you will need to load the Microsoft Excel Analysis ToolPak add-in. Follow the procedure in Section GS 8.1 before continuing.

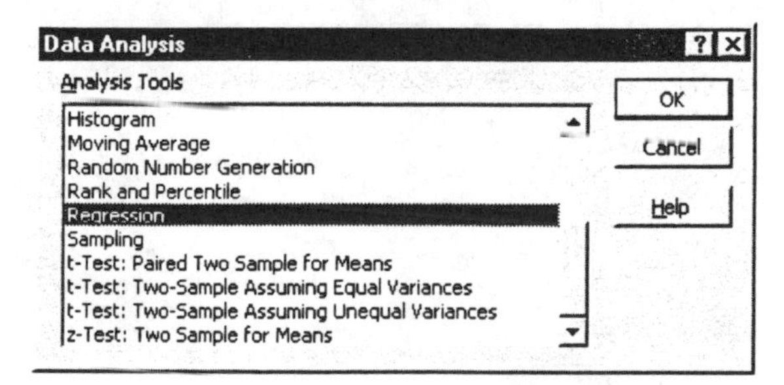

3. Complete the Regression dialog box as shown below. Click **OK**.

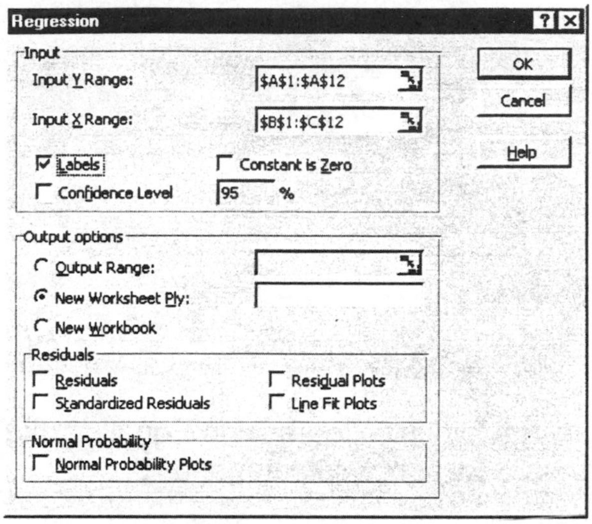

Regression

Input
Input Y Range: $A$1:$A$12
Input X Range: $B$1:$C$12
☑ Labels ☐ Constant is Zero
☐ Confidence Level 95 %

OK
Cancel
Help

Output options
◯ Output Range:
◉ New Worksheet Ply:
◯ New Workbook

Residuals
☐ Residuals ☐ Residual Plots
☐ Standardized Residuals ☐ Line Fit Plots

Normal Probability
☐ Normal Probability Plots

You will want to make some of the columns wider so that you can read all the labels. The coefficient of determination (R^2) is displayed in the upper portion of the output. R^2 is equal to 0.9881. The standard error of estimate is also displayed in the upper portion of the output. The standard error of estimate is equal to 34.1597. The coefficients for the multiple regression equation are displayed in the lower portion of the output under the label "Coefficients."

	A	B	C	D	E	F	G	H	I
1	SUMMARY OUTPUT								
2									
3	*Regression Statistics*								
4	Multiple R	0.9940231							
5	R Square	0.9880819							
6	Adjusted R Square	0.9851024							
7	Standard Error	34.159726							
8	Observations	11							
9									
10	ANOVA								
11		*df*	*SS*	*MS*	*F*	*ignificance F*			
12	Regression	2	773935.4867	386968	331.624	2.02E-08			
13	Residual	8	9335.095087	1166.89					
14	Total	10	783270.5818						
15									
16		*Coefficients*	*Standard Error*	*t Stat*	*P-value*	*Lower 95%*	*Upper 95%*	*Lower 95.0%*	*Upper 95.0%*
17	Intercept	-256.2926	44.46560248	-5.76384	0.000422	-358.831	-153.755	-358.83052	-153.75466
18	sq footage	103.50189	95.23861788	1.08676	0.308798	-116.119	323.1227	-116.11889	323.122683
19	Shopping centers	14.64888	10.9081471	1.34293	0.216148	-10.5054	39.80313	-10.505369	39.8031281

◀

Technology Lab

► Exercise 1 (pg. 457) Constructing a Scatter Plot

1. Open worksheet "Tech9" in the Chapter 9 folder.

2. You will first construct a scatter plot of the tar and nicotine variables. Click on any cell within the table of data. At the top of the screen, click **Insert** and select **Chart** from the menu that appears.

3. In the Chart Type dialog box, select the **XY (Scatter)** chart type by clicking on it. Then click on the topmost chart sub-type. Click **Next**.

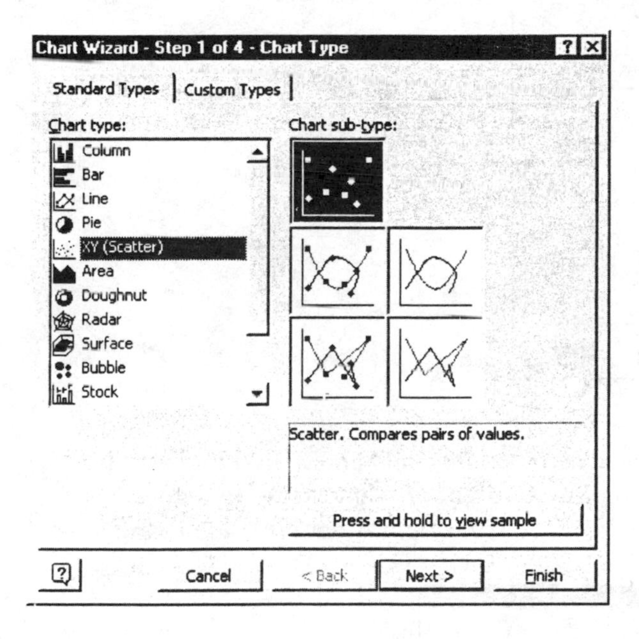

4. In the Chart Source Data dialog box, the entry in the data range field should be **=Tech9!$B$1:$C$23**. Correct the entry and then click **Next**.

Note that the data range includes only columns B and C. These are the columns that contain the tar and nicotine variables that will be included in this scatter plot.

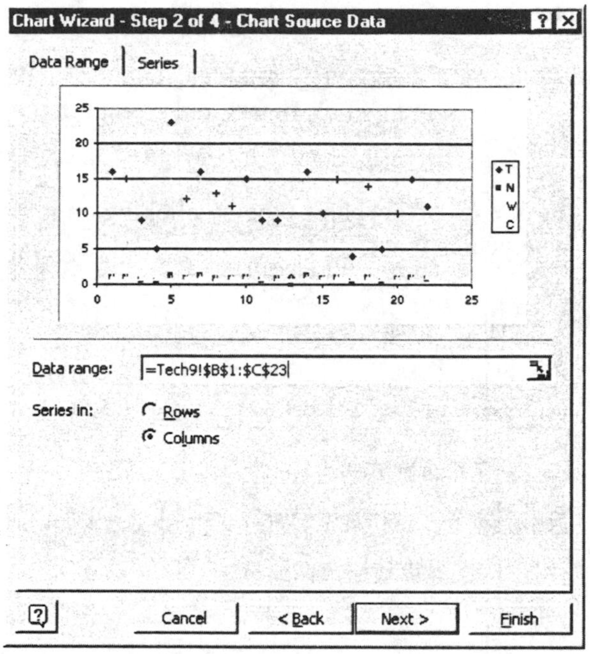

Chart Wizard - Step 2 of 4 - Chart Source Data

Data Range | Series

Data range: =Tech9!$B$1:$C$23

Series in: Rows / Columns

Cancel | < Back | Next > | Finish

5. Click the Titles tab at the top of the Chart Options dialog box. In the Chart title field, change the title to **Tar by Nicotine Content of Cigarettes**. In the Value (X) axis field, enter **Tar in mg**. In the Value (Y) axis field, enter **Nicotine in mg**.

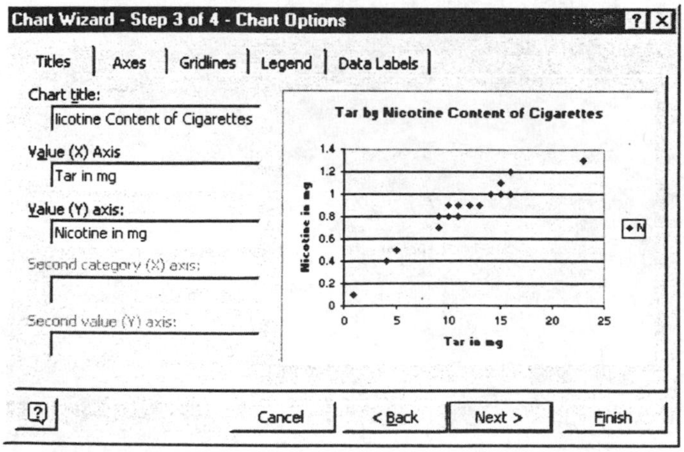

Chart Wizard - Step 3 of 4 - Chart Options

Titles | Axes | Gridlines | Legend | Data Labels

Chart title: licotine Content of Cigarettes

Value (X) Axis: Tar in mg

Value (Y) axis: Nicotine in mg

Second category (X) axis:

Second value (Y) axis:

Tar by Nicotine Content of Cigarettes

Cancel | < Back | Next > | Finish

6. Click the Legend tab at the top of the Chart Options dialog box. Click in the box to the left of Show Legend so tat a checkmark does not appear there. This removes the N legend from the right side of the scatter plot. Click **Next**.

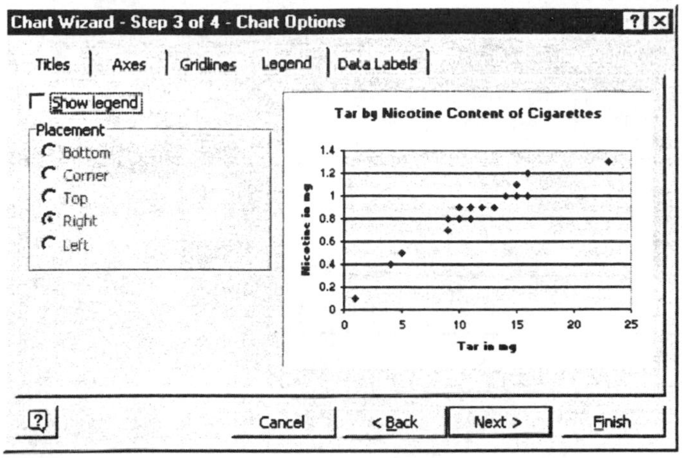

Chart Wizard - Step 3 of 4 - Chart Options

Titles | Axes | Gridlines | Legend | Data Labels

Show legend

Placement
- Bottom
- Corner
- Top
- Right
- Left

Tar by Nicotine Content of Cigarettes

Nicotine in mg

Tar in mg

Cancel | < Back | Next > | Finish

7. The Chart Location dialog box presents two options for placement of the scatter plot. For this example, select **As new sheet**. To do this, click the button to the left of **As new sheet** so that a black dot appears there. Click **Finish**.

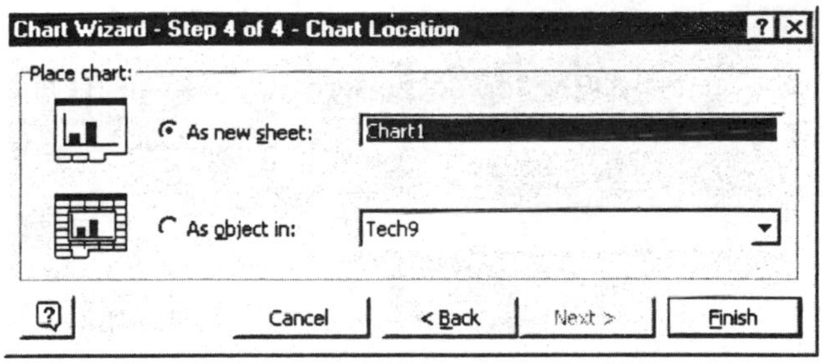

Chart Wizard - Step 4 of 4 - Chart Location

Place chart:

As new sheet: Chart1

As object in: Tech9

Cancel | < Back | Next > | Finish

Your scatter plot should look similar to the scatter plot displayed below.

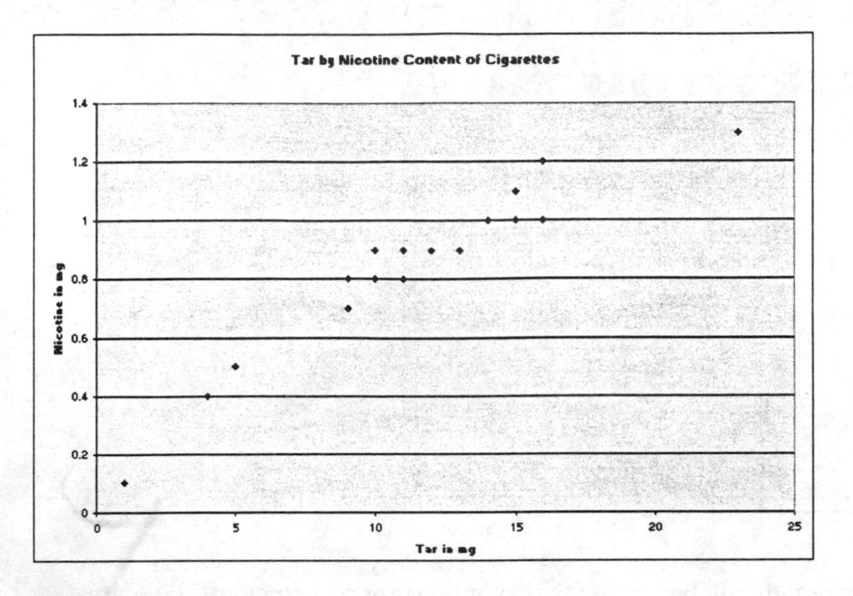

8. You will now construct the scatter plot for the tar and weight variables. Go back to the worksheet that contains the data. To do this, click on the **Tech9** tab near the bottom of the screen.

9. Click on any cell within the table of data. At the top of the screen, click **Insert** and select **Chart** from the menu that appears.

10. In the Chart Type dialog box, select the **XY (Scatter)** chart type by clicking on it. Then click on the topmost chart sub-type. Click **Next**.

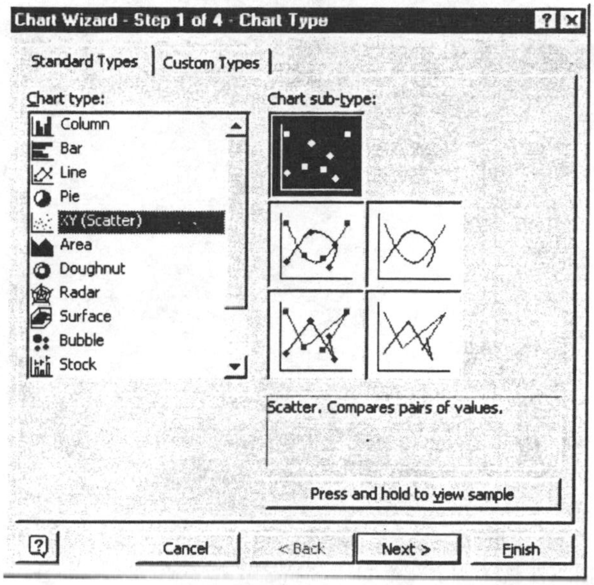

11. Click the **Data Range** tab at the top of the Chart Source Data dialog box. Edit the entry in the data range field so that it reads **=Tech9!$B$1:$D$23**.

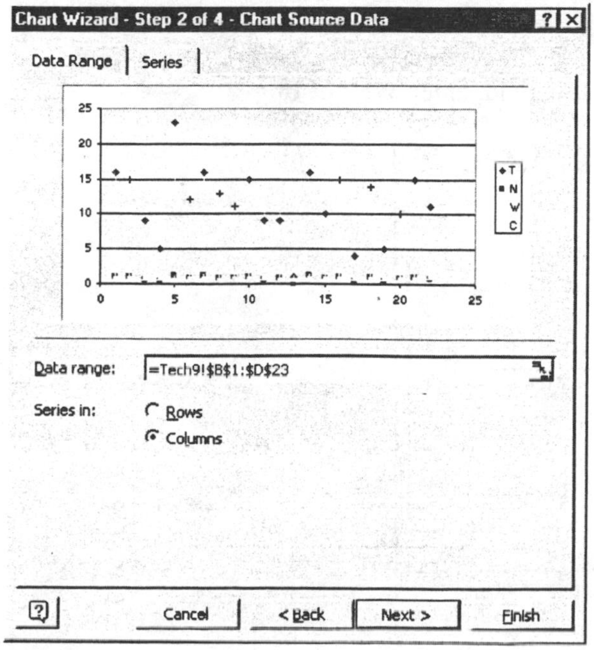

12. Click the **Series** tab at the top of the Chart Source Data dialog box. For Y values, you want the weight variable that is in column D of the worksheet. Edit the entry in the Y Values field so that it reads **=Tech9!$D$2:$D$23**. Click **Next**.

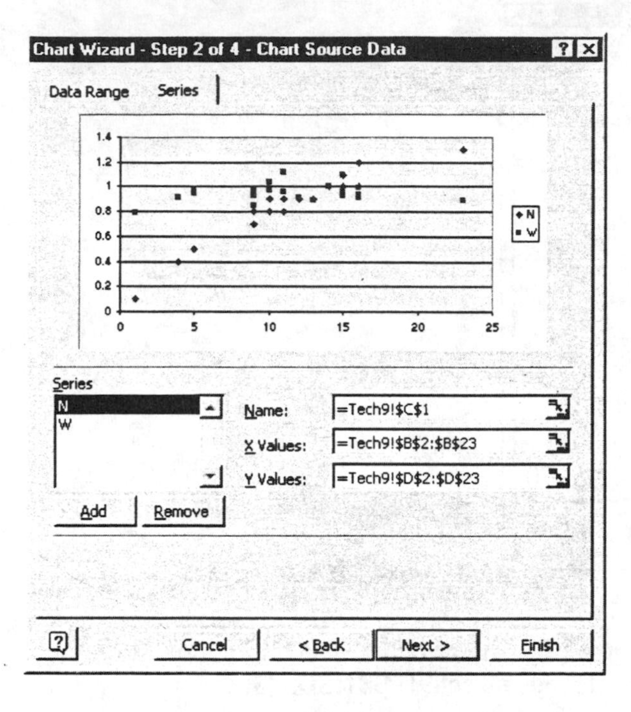

13. Click the Titles tab at the top of the Chart Options dialog box. In the Chart title field, enter **Tar Content by Weight of Cigarettes**. In the Value (X) axis field, enter **Tar in mg**. In the Value (Y) axis field, enter **Weight in g**.

Chart Wizard - Step 3 of 4 - Chart Options

Titles | Axes | Gridlines | Legend | Data Labels

Chart title:
ent by Weight of Cigarettes

Value (X) Axis
Tar in mg

Value (Y) axis:
Weight in g

Second category (X) axis:

Second value (Y) axis:

Tar Content by Weight of Cigarettes
Weight in g
Tar in mg
N
W

Cancel | < Back | Next > | Finish

14. Click the **Legend** tab at the top of the Chart Options dialog box. Click in the box to the left of **Show Legend** so that a checkmark does not appear there. This removes the N and W legends from the right side of the scatter plot. Click **Next**.

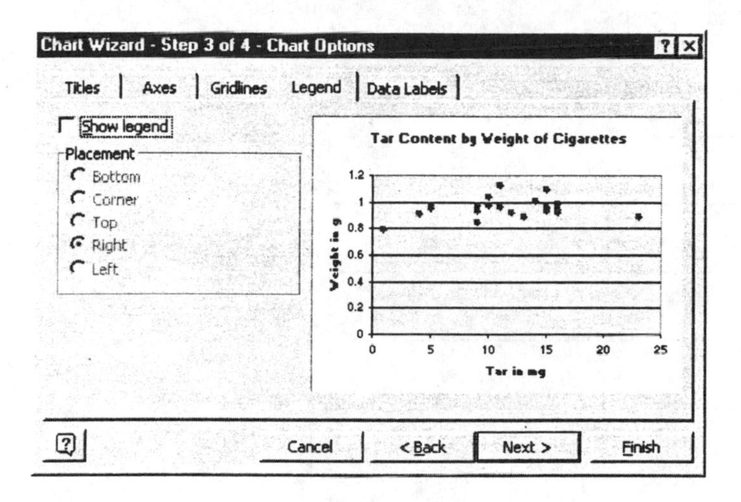

15. The Chart Location dialog box presents two options for placement of the scatter plot. Select **As new sheet**. Click **Finish**.

Chart Wizard - Step 4 of 4 - Chart Location
Place chart:
As new sheet: Chart2
As object in: Tech9
Cancel
< Back
Next >
Finish

16. There is a great deal of unused space in the lower half of this scatter plot because the minimum weight value is 0.79. Instead of starting Y axis values with 0, start with 0.7. To do this, right click on any value along the right axis and select **Format Axis** from the shortcut menu that appears.

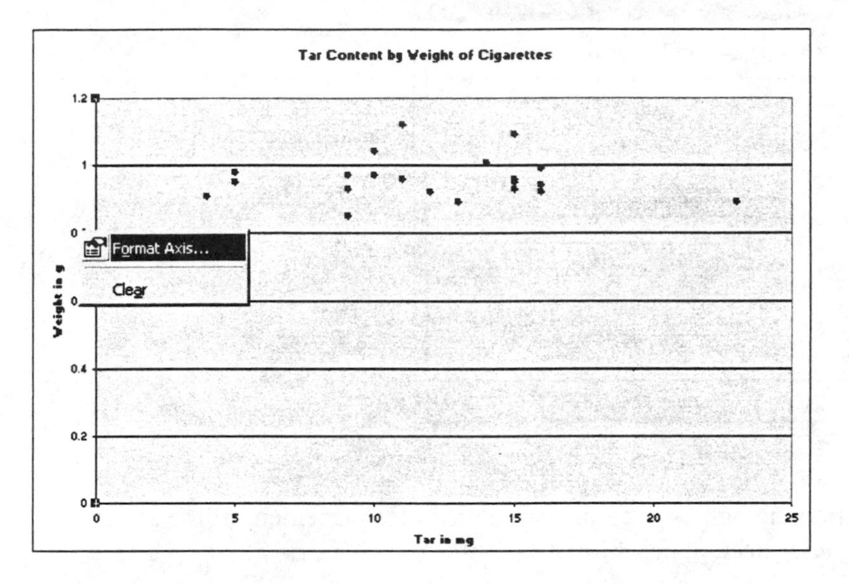

17. Click the **Scale** tab at the top of the Format Axis dialog box. Edit the entry in the Minimum window so that it reads **.7**. Click **OK**.

Format Axis

Patterns | Scale | Font | Number | Alignment

Value (Y) axis scale

Auto

Minimum: .7

Maximum: 1.2

Major unit: 0.2

Minor unit: 0.04

Value (X) axis

Crosses at: 0

Logarithmic scale

Values in reverse order

Value (X) axis crosses at maximum value

OK | Cancel

Your scatter plot should now look similar to the scatter plot shown below.

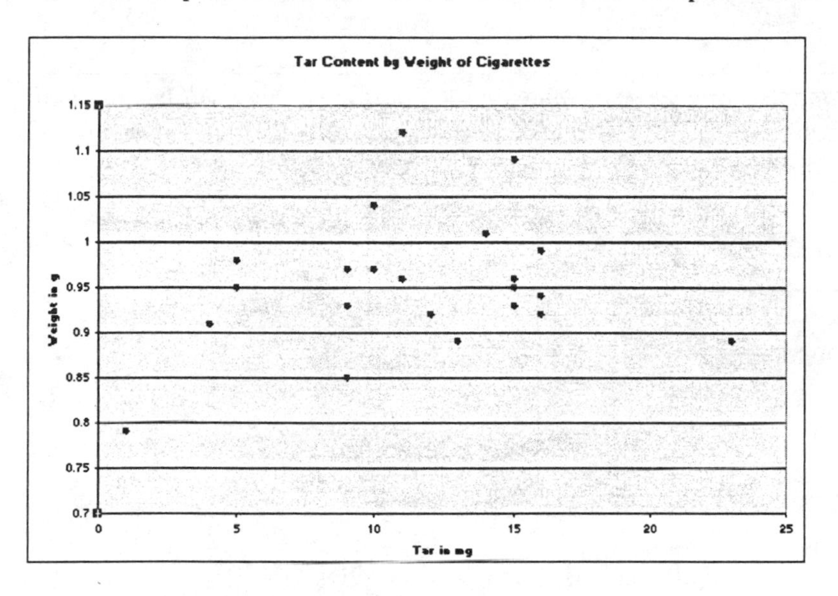

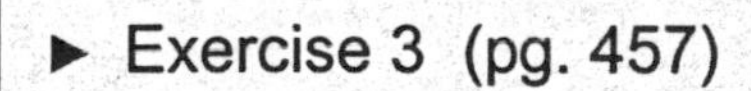

► Exercise 3 (pg. 457) Finding the Correlation Coefficient for Each Pair of Variables

1. Open worksheet "Tech9" in the Chapter 9 folder.

If you have just completed Exercise 1 on page 457 and have not closed the Excel worksheet, return to the sheet containing the data by clicking on the ***Tech9*** *tab at the bottom of the screen.*

2. At the top of the screen, click **Tools → Data Analysis**. Select **Correlation** and click **OK**.

If Data Analysis does not appear as a choice in the Tools menu, you will need to load the Microsoft Excel Analysis ToolPak add-in. Follow the procedure in Section GS 8.1 before continuing.

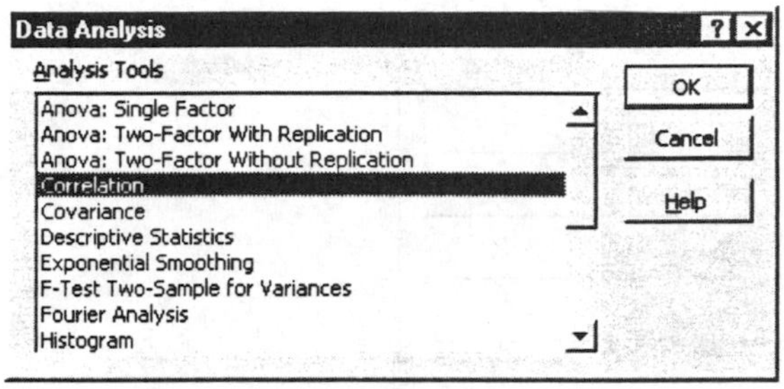

Data Analysis
Analysis Tools
Anova: Single Factor
Anova: Two-Factor With Replication
Anova: Two-Factor Without Replication
Correlation
Covariance
Descriptive Statistics
Exponential Smoothing
F-Test Two-Sample for Variances
Fourier Analysis
Histogram
OK
Cancel
Help

3. Complete the Correlation dialog box as shown below. Click **OK**.

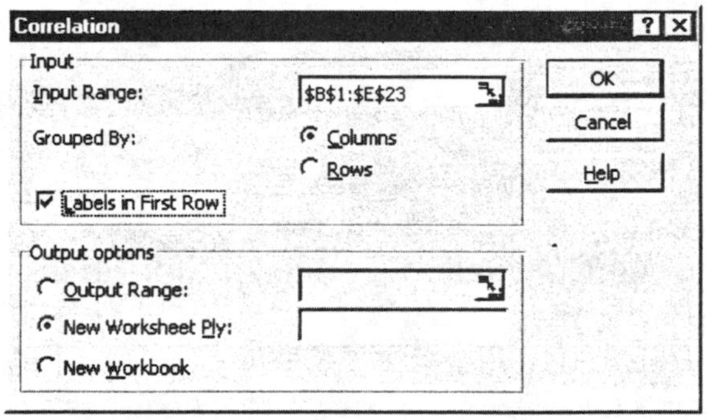

Correlation
Input
Input Range: $B$1:$E$23
Grouped By: Columns / Rows
Labels in First Row
Output options
Output Range:
New Worksheet Ply:
New Workbook
OK
Cancel
Help

The output is a correlation matrix that displays the correlation coefficients for all pairs of variables.

	A	B	C	D	E
1		*T*	*N*	*W*	*C*
2	T	1			
3	N	0.958782	1		
4	W	0.208178	0.298418	1	
5	C	0.91021	0.934994	0.361635	1

◀

► Exercise 4 (pg. 457) — Finding the Regression Line

1. Open worksheet “Tech9” in the Chapter 9 folder.

*If you have just completed Exercise 1 or Exercise 3 on page 457 and have not closed the Excel worksheet, return to the sheet containing the data by clicking on the **Tech9** tab at the bottom of the screen.*

2. You will first find the regression line equation for the tar and nicotine variables. Tar will be the Y variable in the equation and nicotine will be the X variable. At the top of the screen, click **Tools → Data Analysis**. Select **Regression** and click **OK**.

If Data Analysis does not appear as a choice in the Tools menu, you will need to load the Microsoft Excel Analysis ToolPak add-in. Follow the procedure in Section GS 8.1 before continuing.

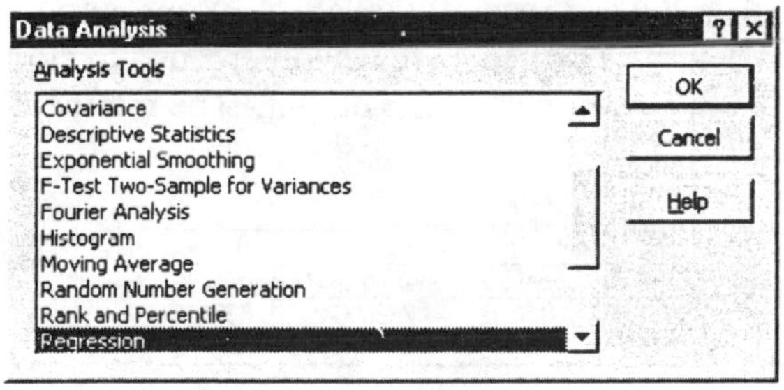

Data Analysis
Analysis Tools
Covariance
Descriptive Statistics
Exponential Smoothing
F-Test Two-Sample for Variances
Fourier Analysis
Histogram
Moving Average
Random Number Generation
Rank and Percentile
Regression
OK
Cancel
Help

3. Complete the Regression dialog box as shown below. Click **OK**.

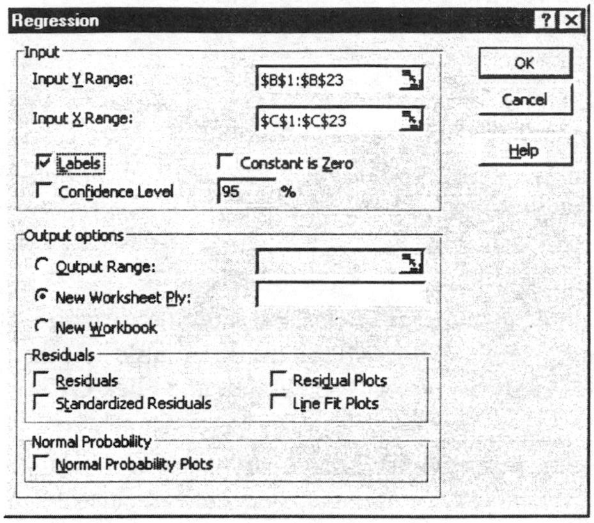

Regression

Input
Input Y Range: $B$1:$B$23
Input X Range: $C$1:$C$23
☑ Labels ☐ Constant is Zero
☐ Confidence Level 95 %

Output options
○ Output Range:
◉ New Worksheet Ply:
○ New Workbook

Residuals
☐ Residuals ☐ Residual Plots
☐ Standardized Residuals ☐ Line Fit Plots

Normal Probability
☐ Normal Probability Plots

OK Cancel Help

Make the columns in the output wider so that you can read all the labels. Your output should appear similar to the output shown below. The values for the intercept and slope of the regression line equation are shown in the lower part of the output. The intercept is –2.7149 and the slope is 16.6876.

	A	B	C	D	E	F	G	H	I
16		*Coefficients*	*Standard Error*	*t Stat*	*P-value*	*Lower 95%*	*Upper 95%*	*Lower 95.0%*	*Upper 95.0%*
17	Intercept	-2.714885	0.994704233	-2.72934	0.01292	-4.7898004	-0.63997	-4.7898004	-0.639969
18	N	16.687631	1.105847665	15.0904	2.15E-12	14.3808743	18.99439	14.3808743	18.9943878

4. You will now find the regression line equation for the tar and carbon monoxide variables. Tar will be the Y variable in the equation and carbon monoxide will be the X variable. Go back to the sheet containing the data by clicking on the **Tech9** tab at the bottom of the screen. Then, at the top of the screen, click **Tools → Data Analysis**. Select **Regression** and click **OK**.

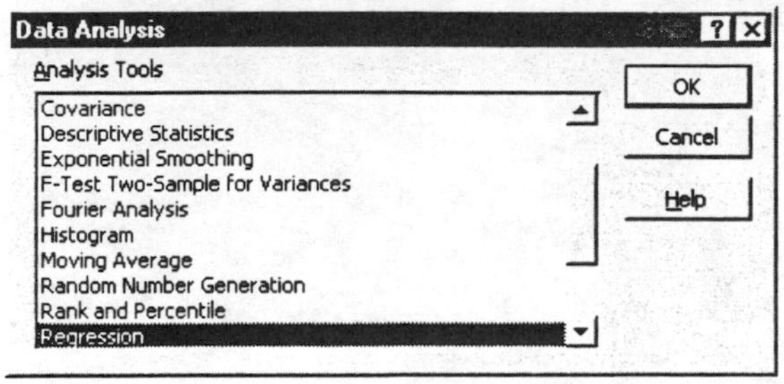

Data Analysis

Analysis Tools
Covariance
Descriptive Statistics
Exponential Smoothing
F-Test Two-Sample for Variances
Fourier Analysis
Histogram
Moving Average
Random Number Generation
Rank and Percentile
Regression

OK Cancel Help

5. Complete the Regression dialog box as shown below. Click **OK**.

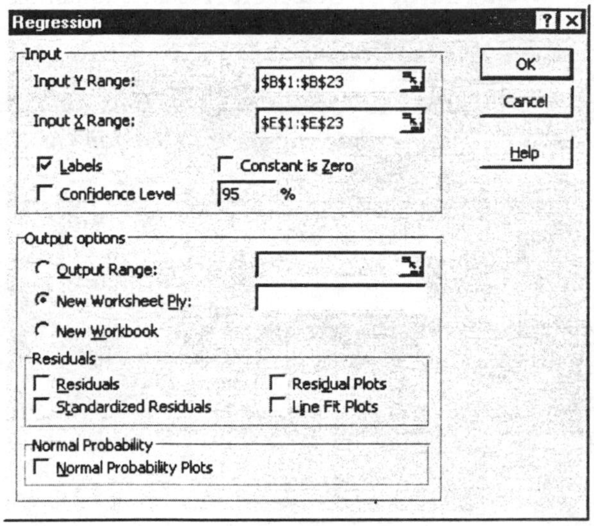

Regression

Input

Input Y Range: $B$1:$B$23

Input X Range: $E$1:$E$23

☑ Labels ☐ Constant is Zero

☐ Confidence Level 95 %

OK

Cancel

Help

Output options

○ Output Range:

◉ New Worksheet Ply:

○ New Workbook

Residuals

☐ Residuals ☐ Residual Plots

☐ Standardized Residuals ☐ Line Fit Plots

Normal Probability

☐ Normal Probability Plots

6. Make the columns in the output wider so that you can read all the labels. Your output should appear similar to the output shown below. The values for the intercept and slope of the regression line equation are shown in the lower part of the output. The intercept is –1.3882 and the slope is 1.1291.

	A	B	C	D	E	F	G	H	I
16		Coefficients	Standard Error	t Stat	P-value	Lower 95%	Upper 95%	Lower 95.0%	Upper 95.0%
17	Intercept	-1.388179	1.391541919	-0.99758	0.330399	-4.29088	1.514525	-4.2908831	1.51452532
18	C	1.1291267	0.114879059	9.82883	4.23E-09	0.889493	1.36876	0.88949332	1.36876014

◀

► Exercise 6 (pg. 457)

Finding Multiple Regression Equations

1. Open worksheet "Tech9" in the Chapter 9 folder.

*If you have just completed Exercise 1, Exercise 3, or Exercise 4 on page 457 and have not closed the Excel worksheet, return to the sheet containing the data by clicking on the **Tech9** tab at the bottom of the screen.*

2. At the top of the screen, click **Tools** → **Data Analysis**. Select **Regression** and click **OK**.

If Data Analysis does not appear as a choice in the Tools menu, you will need to load the Microsoft Excel Analysis ToolPak add-in. Follow the procedure in Section GS 8.1 before continuing.

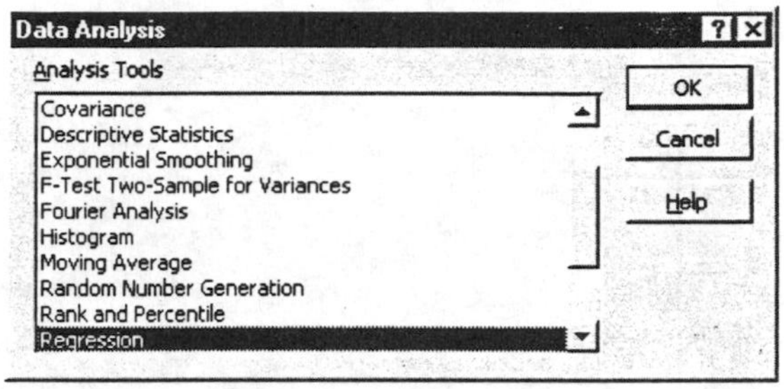

Data Analysis

Analysis Tools

- Covariance
- Descriptive Statistics
- Exponential Smoothing
- F-Test Two-Sample for Variances
- Fourier Analysis
- Histogram
- Moving Average
- Random Number Generation
- Rank and Percentile
- Regression

OK | Cancel | Help

3. You will first construct the multiple regression equation using three predictor variables—nicotine, weight and carbon monoxide. Complete the Regression dialog box as shown below. Click **OK**.

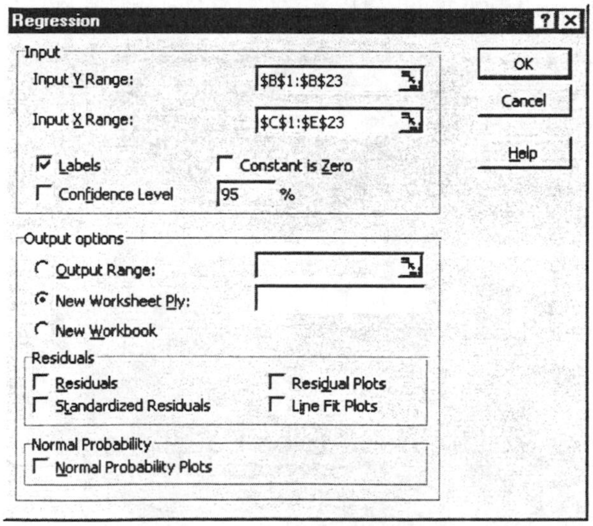

Regression

Input
- Input Y Range: $B$1:$B$23
- Input X Range: $C$1:$E$23
- [x] Labels
- [ ] Constant is Zero
- [ ] Confidence Level 95 %

Output options
- ( ) Output Range:
- (•) New Worksheet Ply:
- ( ) New Workbook

Residuals
- [ ] Residuals
- [ ] Residual Plots
- [ ] Standardized Residuals
- [ ] Line Fit Plots

Normal Probability
- [ ] Normal Probability Plots

OK | Cancel | Help

You will want to make some of the columns wider so that you can read all the labels. Your output should look similar to the output shown below. In the intercept and coefficients of the predictor variables are displayed in the bottom four lines of the output under the label "Coefficients."

	A	B	C	D	E	F	G	H	I
16		*Coefficients*	*Standard Error*	*t Stat*	*P-value*	*Lower 95%*	*Upper 95%*	*Lower 95.0%*	*Upper 95.0%*
17	Intercept	3.4469686	4.2381019	0.813328	0.426658	-5.45696	12.3509	-5.45696	12.3508972
18	N	14.350349	3.091521395	4.64184	0.000203	7.855298	20.8454	7.8552981	20.845399
19	W	-6.996539	4.656584728	-1.5025	0.150308	-16.7797	2.78659	-16.779668	2.78659032
20	C	0.2183635	0.225569488	0.968054	0.345846	-0.25554	0.692268	-0.2555408	0.69226773

4. Now you will construct the regression equation using two predictor variables—nicotine and carbon monoxide. Return to the worksheet containing the data by clicking on the **Tech9** tab at the bottom of the screen.

5. The predictor variables must be located in adjacent columns in the Excel worksheet. So you'll begin by making a copy of the worksheet and then deleting the column of data between nicotine and carbon monoxide. At the top of the screen, click **Edit → Move or Copy Sheet.**

6. In the Move or Copy dialog box, click in the box to the left of **Create a copy** to place a checkmark there. Click **OK.**

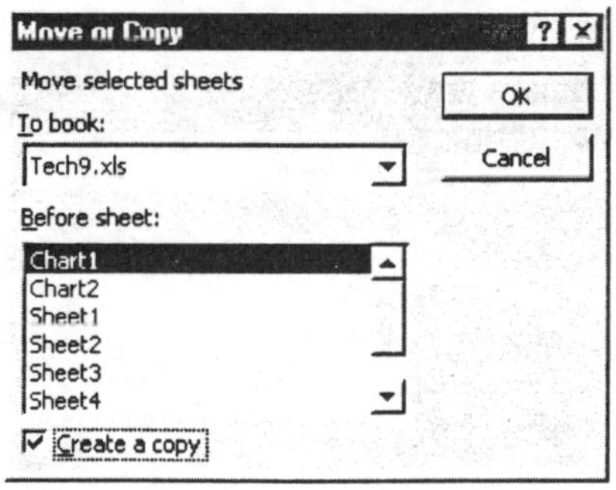

Move or Copy
Move selected sheets
To book:
Tech9.xls
Before sheet:
Chart1
Chart2
Sheet1
Sheet2
Sheet3
Sheet4
Create a copy
OK
Cancel

7. The copy is called "Tech9(2)." To delete the column with the weight data, first click directly on D in the letter row at the top of the worksheet. Column D will be highlighted. Then, at the top of the screen, click **Edit → Delete**. Your worksheet should now look like the one shown below.

	A	B	C	D	E	F	G
1	Brand	T	N	C			
2	Alpine	16	1	15			
3	Benson &	15	1.1	15			
4	Camel	9	0.7	11			
5	Carlton	5	0.5	3			
6	Chesterfiel	23	1.3	15			
7	Kent	12	0.9	12			

8. At the top of the screen, click **Tools → Data Analysis**. Select **Regression** and click **OK**.

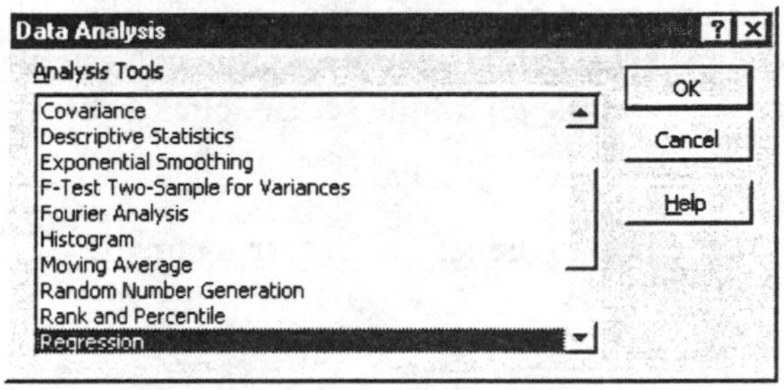

Data Analysis

Analysis Tools

- Covariance
- Descriptive Statistics
- Exponential Smoothing
- F-Test Two-Sample for Variances
- Fourier Analysis
- Histogram
- Moving Average
- Random Number Generation
- Rank and Percentile
- Regression

OK | Cancel | Help

9. Complete the Regression dialog box as shown below. Click **OK**.

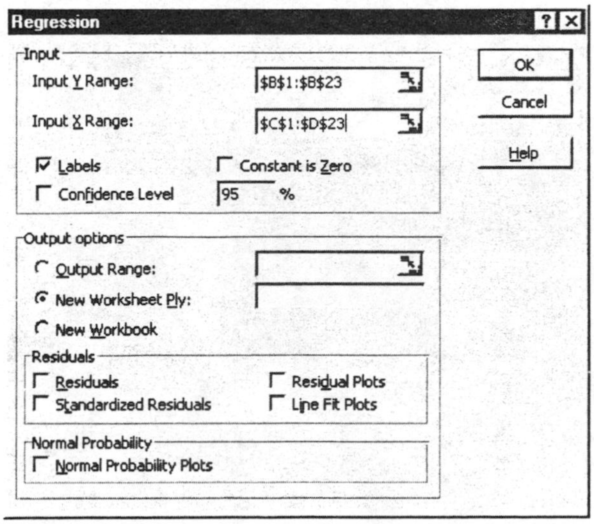

Regression

Input

Input Y Range: $B$1:$B$23

Input X Range: $C$1:$D$23

☑ Labels ☐ Constant is Zero

☐ Confidence Level 95 %

Output options

○ Output Range:

◉ New Worksheet Ply:

○ New Workbook

Residuals

☐ Residuals ☐ Residual Plots

☐ Standardized Residuals ☐ Line Fit Plots

Normal Probability

☐ Normal Probability Plots

OK | Cancel | Help

After making some of the columns wider, your output should appear similar to the output shown below. The intercept and coefficients of the predictor variables are displayed in the bottom lines of the output under the label "Coefficients."

	A	B	C	D	E	F	G	H	I
16		*Coefficients*	*Standard Error*	*t Stat*	*P-value*	*Lower 95%*	*Upper 95%*	*Lower 95.0%*	*Upper 95.0%*
17	Intercept	-2.748014	1.012498867	-2.71409	0.013764	-4.8672	-0.62883	-4.8671992	-0.6288289
18	N	14.908175	3.16908666	4.70425	0.000154	8.275198	21.54115	8.2751979	21.5411513
19	C	0.1356453	0.225871018	0.600543	0.555241	-0.33711	0.608399	-0.3371083	0.60839896

◀

Chi-Square Tests and the F-Distribution

CHAPTER 10

Section 10.2

▶ Example 3 (pg. 481)

Chi-Square Independence Test

If the PHStat add-in has not been loaded, you will need to load it before continuing. Follow the instructions in Section GS 8.2.

1. First, open a new Excel worksheet. Then, at the top of the screen, select **PHStat → c-Sample Tests → Chi-Square Test**.

2. Complete the Chi-Square Test dialog box as shown below. Click **OK**.

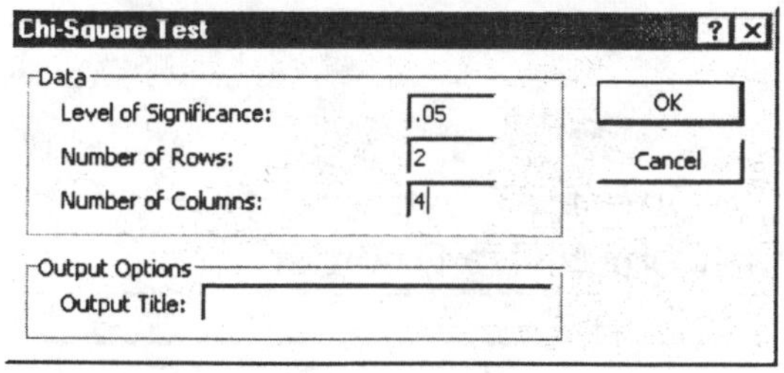

Chi-Square Test

Data

Level of Significance: .05

Number of Rows: 2

Number of Columns: 4

OK

Cancel

Output Options

Output Title:

3. A Chi-Square Test worksheet will appear. Complete the Observed Frequencies table in this worksheet as shown below. You will notice that values in other locations of the worksheet change as you enter the observed frequencies.

	A	B	C	D	E	F	G	H
1	Chi-Square Test							
2								
3	Observed Frequencies:				Column variable			
4			Row variable	C1	C2	C3	C4	Total
5			R1	40	53	26	6	125
6			R2	34	68	37	11	150
7			Total	74	121	63	17	275

The critical value, the obtained chi-square test statistic, and the p-value are displayed in rows 20 through 22. The statistical decision is displayed in row 23.

	A	B
16	Level of Significance	0.05
17	Number of Rows	2
18	Number of Columns	4
19	Degrees of Freedom	3
20	Critical Value	7.814725
21	Chi-Square Test Statistic	3.493356
22	p-Value	0.321625
23	Do not reject the null hypothesis	

◀

▶ Exercise 5 (pg. 483) Testing the Drug for Treatment of Obsessive-Compulsive Disorder

If the PHStat add-in has not been loaded, you will need to load it before continuing. Follow the instructions in Section GS 8.2.

1. Open a new Excel worksheet. At the top of the screen, select **PHStat → c-Sample Tests → Chi-Square Test.**

2. Complete the Chi-Square Test dialog box as shown below. Click **OK**.

Chi-Square Test	
Data	
Level of Significance:	.10
Number of Rows:	2
Number of Columns:	2
Output Options	
Output Title:	

OK | Cancel

4. A Chi-Square Test worksheet will appear. Complete the Observed Frequencies table in this worksheet as shown below.

	A	B	C	D	E	F
1	**Chi-Square Test**					
2						
3	**Observed Frequencies:**			**Column variable**		
4			**Row variable**	**C1**	**C2**	**Total**
5			**R1**	39	25	64
6			**R2**	54	70	124
7			**Total**	93	95	188

The critical value, the obtained chi-square test statistic, and the p-value are displayed in rows 20 through 22. The statistical decision is displayed in row 23.

	A	B
16	**Level of Significance**	**0.1**
17	Number of Rows	2
18	Number of Columns	2
19	Degrees of Freedom	1
20	**Critical Value**	**2.705541**
21	**Chi-Square Test Statistic**	**5.106328**
22	***p*-Value**	**0.023839**
23	**Reject the null hypothesis**	

◀

Section 10.3

► Example 3 (pg. 491) Performing a Two-Sample F-Test

If the PHStat add-in has not been loaded, you will need to load it before continuing. Follow the instructions in Section GS 8.2.

1. First open a new Excel worksheet. Then, at the top of the screen, select **PHStat → Two-Sample Tests → F Test for Differences in Two Variances.**

2. Complete the F Test for Differences in Two Variances dialog box as shown below. Click **OK**.

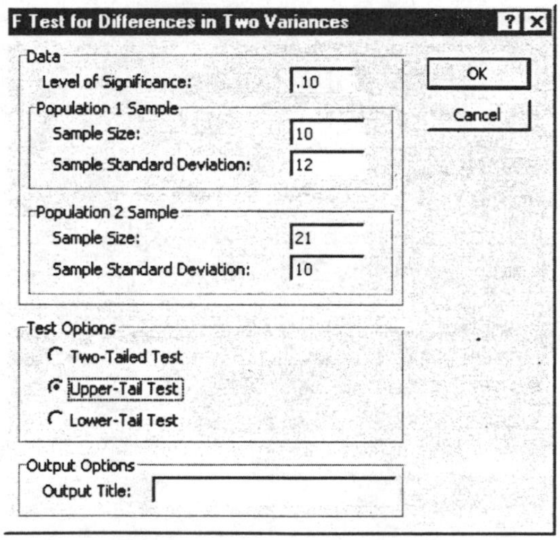

F Test for Differences in Two Variances

Data
Level of Significance: .10
Population 1 Sample
Sample Size: 10
Sample Standard Deviation: 12
Population 2 Sample
Sample Size: 21
Sample Standard Deviation: 10

Test Options
Two-Tailed Test
Upper-Tail Test
Lower-Tail Test

Output Options
Output Title:

OK
Cancel

Your output should appear similar to the output displayed below.

	A	B	C	D	E
1	**F Test for Differences in Two Variances**				
2					
3	**Level of Significance**	**0.1**			
4	**Population 1 Sample**				
5	**Sample Size**	**10**			
6	**Sample Standard Deviation**	**12**			
7	**Population 2 Sample**				
8	**Sample Size**	**21**			
9	**Sample Standard Deviation**	**10**			
10	*F*-Test Statistic	1.44			
11	Population 1 Sample Degrees of Freedom	9			
12	Population 2 Sample Degrees of Freedom	20			
13				Calculations Area	
14	**Upper-Tail Test**			FDIST value	0.2369
15	**Upper Critical Value**	**1.964853**		1-FDIST value	0.7631
16	***p*-Value**	**0.2369**			
17	**Do not reject the null hypothesis**				

◀

Section 10.4

▸ Example 2 (pg. 501)

Performing an ANOVA Test

1. Open worksheet "Airline" in the Chapter 10 folder.

2. At the top of the screen, click **Tools → Data Analysis**. Select **ANOVA: Single Factor** and click **OK**.

If Data Analysis does not appear as a choice in the Tools menu, you will need to load the Microsoft Excel Analysis ToolPak add-in. Follow the procedure in Section GS 8.1 before continuing.

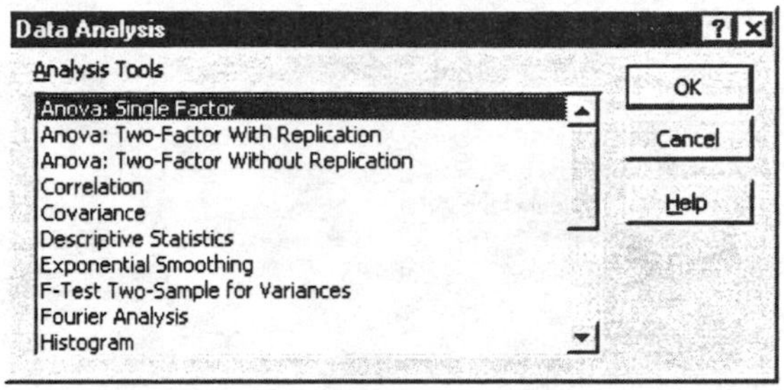

Data Analysis
Analysis Tools
Anova: Single Factor
Anova: Two-Factor With Replication
Anova: Two-Factor Without Replication
Correlation
Covariance
Descriptive Statistics
Exponential Smoothing
F-Test Two-Sample for Variances
Fourier Analysis
Histogram
OK
Cancel
Help

3. Complete the ANOVA: Single Factor dialog box as shown below. Click **OK**.

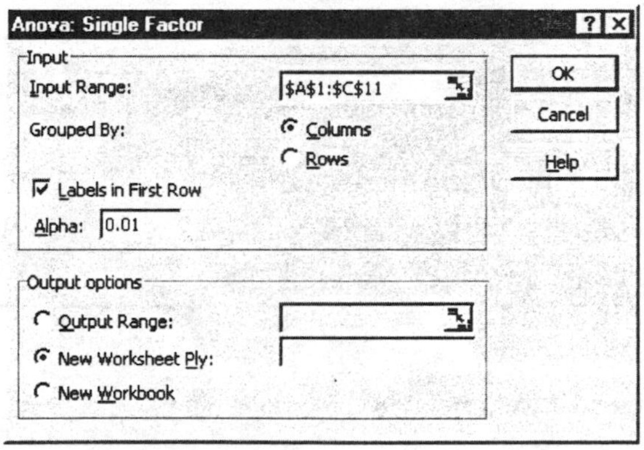

Anova: Single Factor
Input
Input Range: $A$1:$C$11
Grouped By: Columns / Rows
Labels in First Row
Alpha: 0.01
Output options
Output Range:
New Worksheet Ply:
New Workbook
OK
Cancel
Help

Make column A wider so that you can read all the labels. Your output should appear similar to the output displayed below.

	A	B	C	D	E	F	G
1	Anova: Single Factor						
2							
3	SUMMARY						
4	*Groups*	*Count*	*Sum*	*Average*	*Variance*		
5	Airline1	10	1238	123.8	106.6222		
6	Airline2	10	1315	131.5	120.2778		
7	Airline3	10	1427	142.7	215.1222		
8							
9							
10	ANOVA						
11	*Source of Variation*	*SS*	*df*	*MS*	*F*	*P-value*	*F crit*
12	Between Groups	1806.467	2	903.2333	6.130235	0.006383	5.488118
13	Within Groups	3978.2	27	147.3407			
14							
15	Total	5784.667	29				

◀

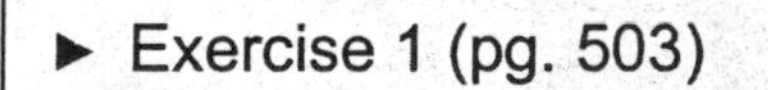

Testing the Claim That the Mean Costs Per Month are Different

1. Open worksheet "Ex10_4-1" in the Chapter 10 folder.

2. At the top of the screen, click **Tools → Data Analysis**. Select **ANOVA: Single Factor** and click **OK**.

If Data Analysis does not appear as a choice in the Tools menu, you will need to load the Microsoft Excel Analysis ToolPak add-in. Follow the procedure in Section GS 8.1 before continuing.

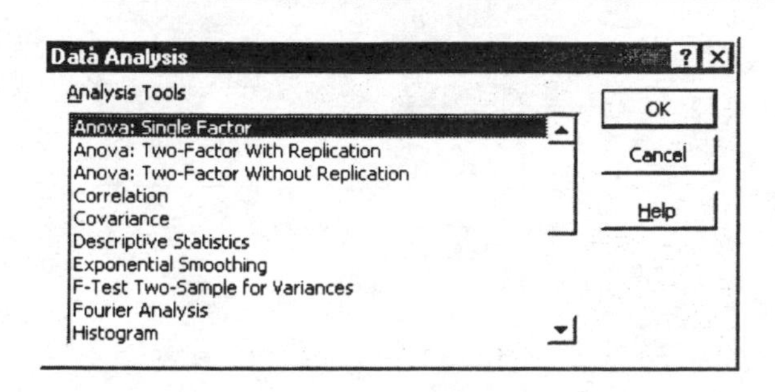

3. Complete the ANOVA: Single Factor dialog box as shown below. Click **OK**.

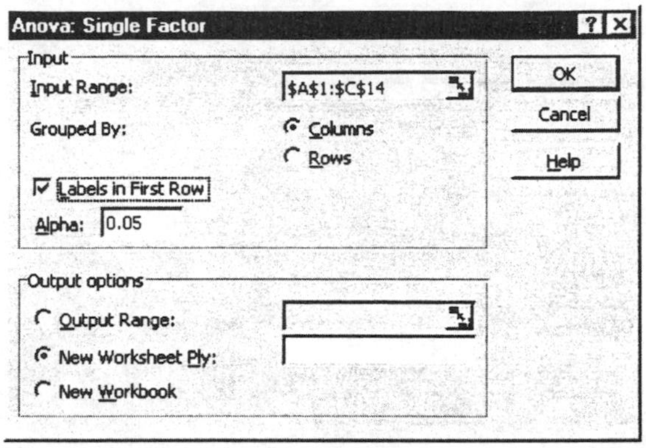

Anova: Single Factor
Input
Input Range: $A$1:$C$14
Grouped By: Columns / Rows
Labels in First Row
Alpha: 0.05
Output options
Output Range:
New Worksheet Ply:
New Workbook
OK
Cancel
Help

Make column A wider so that you can read all the labels. Your output should appear similar to the output displayed below.

	A	B	C	D	E	F	G
1	Anova: Single Factor						
2							
3	SUMMARY						
4	*Groups*	*Count*	*Sum*	*Average*	*Variance*		
5	Moderate	13	15.49	1.191538	0.677964		
6	Low	13	12.52	0.963077	0.145173		
7	Very Low	4	6.75	1.6875	2.530358		
8							
9							
10	ANOVA						
11	*Source of Variation*	*SS*	*df*	*MS*	*F*	*P-value*	*F crit*
12	Between Groups	1.630026	2	0.815013	1.2597	0.299888	3.354131
13	Within Groups	17.46872	27	0.64699			
14							
15	Total	19.09875	29				

Technology Lab

▶ Exercise 2 (pg. 508) Determining Whether the Populations Have Equal Variances

If the PHStat add-in has not been loaded, you will need to load it before continuing. Follow the instructions in Section GS 8.2.

1. Open worksheet "Tech10-a" in the Chapter 10 folder.

2. Because you are working with raw data rather than summary statistics, you first need to calculate the standard deviations of the three groups. At the top of the screen, click **Tools → Data Analysis**. Select **Descriptive Statistics** and click **OK**.

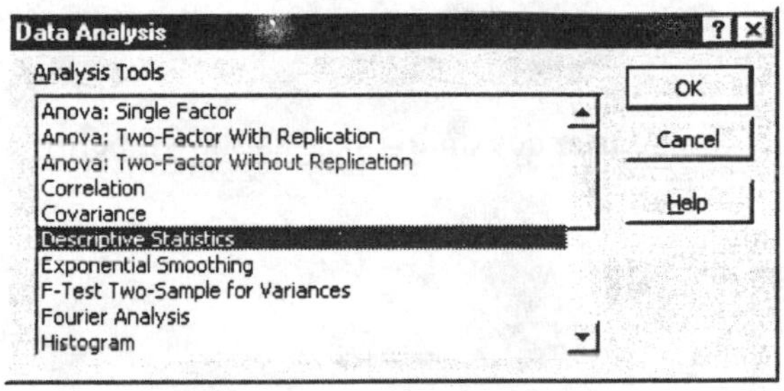

Data Analysis
Analysis Tools
Anova: Single Factor
Anova: Two-Factor With Replication
Anova: Two-Factor Without Replication
Correlation
Covariance
Descriptive Statistics
Exponential Smoothing
F-Test Two-Sample for Variances
Fourier Analysis
Histogram
OK
Cancel
Help

3. Complete the Descriptive Statistics dialog box as shown below. Click **OK**.

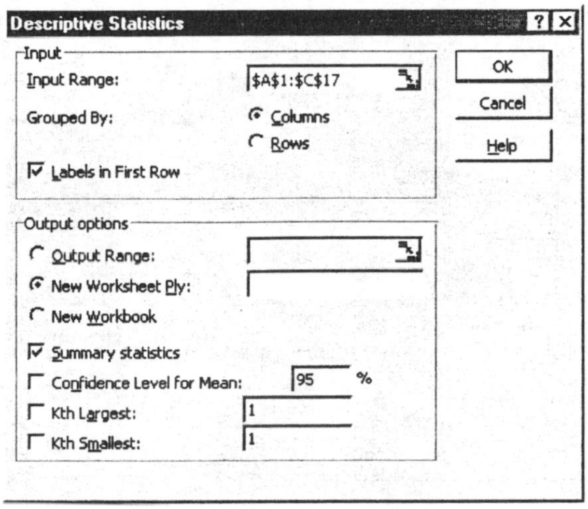

Descriptive Statistics
Input
Input Range: $A$1:$C$17
Grouped By: Columns / Rows
Labels in First Row
Output options
Output Range:
New Worksheet Ply:
New Workbook
Summary statistics
Confidence Level for Mean: 95 %
Kth Largest: 1
Kth Smallest: 1
OK
Cancel
Help

4. Make the columns wider so that you can read the labels. The standard deviations of medium, heavy, and multiple purpose vehicles are 436.46, 448.43, and 566.66, respectively. For this exercise, instructions are provided only for testing the equality of the medium and heavy vehicles. At the top of the screen, select **PHStat → Two-Sample Tests → F Test for Differences in Two Variances.**

	A	B	C	D	E	F
1	*Medium*		*Heavy*		*Multiple*	
2						
3	Mean	1042.75	Mean	947.9375	Mean	971.125
4	Standard Error	109.116	Standard Error	112.1066	Standard Error	141.6639
5	Median	978.5	Median	872.5	Median	966.5
6	Mode	#N/A	Mode	#N/A	Mode	430
7	Standard Deviation	436.4639	Standard Deviation	448.4264	Standard Deviation	566.6558
8	Sample Variance	190500.7	Sample Variance	201086.2	Sample Variance	321098.8
9	Kurtosis	-1.43386	Kurtosis	-0.92956	Kurtosis	-1.24662
10	Skewness	0.001657	Skewness	0.107502	Skewness	0.384132
11	Range	1325	Range	1468	Range	1644
12	Minimum	332	Minimum	260	Minimum	267
13	Maximum	1657	Maximum	1728	Maximum	1911
14	Sum	16684	Sum	15167	Sum	15538
15	Count	16	Count	16	Count	16

5. Complete the F Test for Differences in Two Variances dialog box as shown below. Click **OK**.

F Test for Differences in Two Variances

Data
Level of Significance: 0.05
Population 1 Sample
Sample Size: 16
Sample Standard Deviation: 436.46
Population 2 Sample
Sample Size: 16
Sample Standard Deviation: 448.43
Test Options
(•) Two-Tailed Test
( ) Upper-Tail Test
( ) Lower-Tail Test
Output Options
Output Title:

OK
Cancel

Your output should appear similar to the output displayed below.

	A	B	C	D	E
1	**F Test for Differences in Two Variances**				
2					
3	**Level of Significance**	**0.05**			
4	**Population 1 Sample**				
5	**Sample Size**	**16**			
6	**Sample Standard Deviation**	**436.46**			
7	**Population 2 Sample**				
8	**Sample Size**	**16**			
9	**Sample Standard Deviation**	**448.43**			
10	*F*-Test Statistic	0.947326			
11	Population 1 Sample Degrees of Freedom	15			
12	Population 2 Sample Degrees of Freedom	15			
13				Calculations Area	
14	**Two-Tailed Test**			FDIST value	0.541038
15	**Lower Critical Value**	**0.349395**		1-FDIST value	0.458962
16	**Upper Critical Value**	**2.862095**			
17	***p*-Value**	**0.917923**			
18	**Do not reject the null hypothesis**				

◀

► Exercise 4 (pg. 508) Testing the Claim That the Three Types of Vehicles Have the Same Injury Potential

1. Open worksheet "Tech10-a" in the Chapter 10 folder.

If you have just completed Exercise 2 on page 508 and have not yet closed the Excel worksheet, return to the sheet with the data by clicking on the ***Tech10-a*** *tab near the bottom of the screen.*

2. At the top of the screen, click **Tools → Data Analysis**. Select **ANOVA: Single Factor** and click **OK**.

If Data Analysis does not appear as a choice in the Tools menu, you will need to load the Microsoft Excel Analysis ToolPak add-in. Follow the procedure in Section GS 8.1 before continuing.

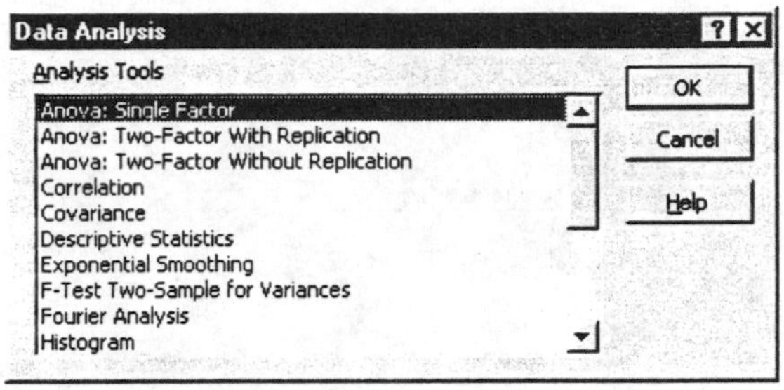

Data Analysis

Analysis Tools

Anova: Single Factor
Anova: Two-Factor With Replication
Anova: Two-Factor Without Replication
Correlation
Covariance
Descriptive Statistics
Exponential Smoothing
F-Test Two-Sample for Variances
Fourier Analysis
Histogram

OK
Cancel
Help

3. Complete the ANOVA: Single Factor dialog box as shown below. Click **OK**.

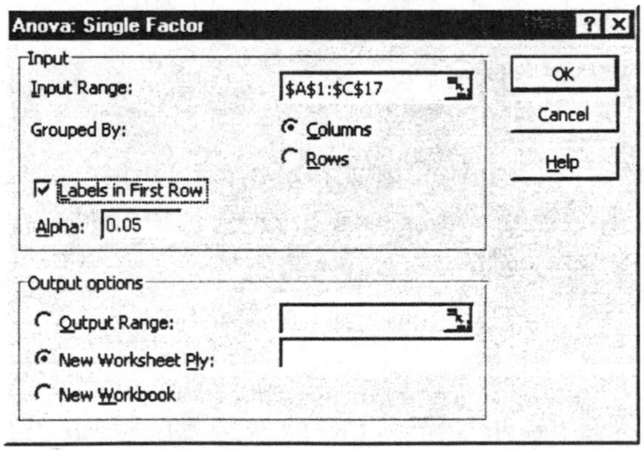

Anova: Single Factor

Input
Input Range: $A$1:$C$17
Grouped By: Columns / Rows
Labels in First Row
Alpha: 0.05

Output options
Output Range:
New Worksheet Ply:
New Workbook

OK
Cancel
Help

Make column A wider so that you can read all the labels. Your output should appear similar to the output shown below.

	A	B	C	D	E	F	G
1	Anova: Single Factor						
2							
3	SUMMARY						
4	*Groups*	*Count*	*Sum*	*Average*	*Variance*		
5	Medium	16	16684	1042.75	190500.7		
6	Heavy	16	15167	947.9375	201086.2		
7	Multiple	16	15538	971.125	321098.8		
8							
9							
10	ANOVA						
11	*Source of Variation*	*SS*	*df*	*MS*	*F*	*P-value*	*F crit*
12	Between Groups	78171.79	2	39085.9	0.164529	0.848801	3.20432
13	Within Groups	10690286	45	237561.9			
14							
15	Total	10768457	47				

Nonparametric Tests

CHAPTER 11

Section 11.2

► Example 2 (pg. 531) **Performing a Wilcoxon Rank Sum Test**

If the PHStat add-in has not been loaded, you will need to load it before continuing. Follow the instructions in Section GS 8.2.

1. Open worksheet "earnings." in the Chapter 11 folder.

2. At the top of the screen, select **PHStat → Two-Sample Tests → Wilcoxon Rank Sum Test**.

3. Complete the Wilcoxon Rank Sum Test dialog box as shown below. The Population 1 Sample Cell Range is A1 through A11. The Population 2 Sample Cell Range is B1 through B13. Click **OK**.

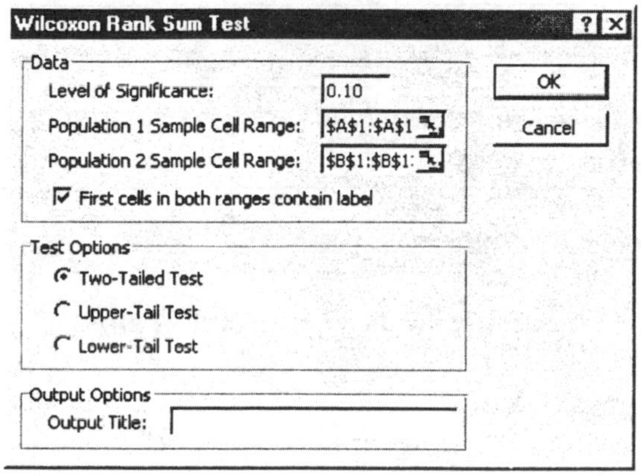

Wilcoxon Rank Sum Test

Data
Level of Significance: 0.10
Population 1 Sample Cell Range: $A$1:$A$1
Population 2 Sample Cell Range: $B$1:$B$1:
First cells in both ranges contain label

OK
Cancel

Test Options
Two-Tailed Test
Upper-Tail Test
Lower-Tail Test

Output Options
Output Title:

Your output should look similar to the output displayed below.

	A	B
1	**Wilcoxon Rank Sum Test**	
2		
3	**Level of Significance**	**0.1**
4	**Population 1 Sample**	
5	**Sample Size**	**10**
6	**Sum of Ranks**	**138**
7	**Population 2 Sample**	
8	**Sample Size**	**12**
9	**Sum of Ranks**	**115**
10		
11	Total Sample Size n	22
12	*T*1 Test Statistic	138
13	*T*1 Mean	115
14	Standard Error of *T*1	15.16575
15	*Z* Test Statistic	1.516575
16		
17	**Two-Tailed Test**	
18	**Lower Critical Value**	**-1.64485**
19	**Upper Critical Value**	**1.644853**
20	***p*-value**	**0.129374**
21	**Do not reject the null hypothesis**	

◄

Section 11.3

► Example 1 (pg. 539) Performing a Kruskal-Wallis Test

If the PHStat add-in has not been loaded, you will need to load it before continuing. Follow the instructions in Section GS 8.2.

1. Open worksheet "payrates" in the Chapter 11 folder.
2. At the top of the screen, select **PHStat → c-Sample Tests → Kruskal-Wallis Rank Test.**

3. Complete the Kruskal-Wallis Rank Test dialog box as shown below. The Sample Data Cell Range is A1 through C11. Click **OK**.

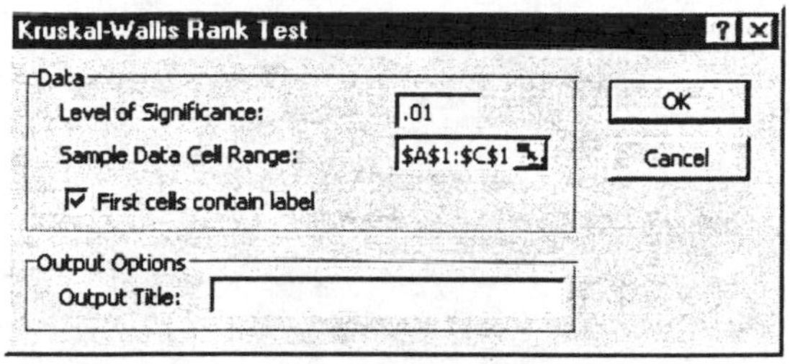

Kruskal-Wallis Rank Test

Data
Level of Significance: .01
Sample Data Cell Range: $A$1:$C$1
☑ First cells contain label

OK
Cancel

Output Options
Output Title:

4. Your output should appear similar to the output shown below. This output is not completely accurate. There are two errors. First, the requested level of significance was 0.01, not 0.05. Second, the critical value of 5.991476 is the critical value for a significance level of 0.05. To correct these errors, simply change the level of significance to **0.01**. As soon as this correction is entered in the worksheet, the critical value in the lower portion of the output also changes.

	A	B
1	**Kruskal Wallis Rank Test**	
2		
3	**Level of Significance**	**0.05**
4	**Group 1**	
5	**Sum of Ranks**	**94.5**
6	**Sample Size**	**10**
7	**Group 2**	
8	**Sum of Ranks**	**223**
9	**Sample Size**	**10**
10	**Group 3**	
11	**Sum of Ranks**	**147.5**
12	**Sample Size**	**10**
13	Sum of Squared Ranks/Sample Size	8041.55
14	Sum of Sample Sizes	30
15	Number of groups	3
16	*H* Test Statistic	10.76194
17	**Critical Value**	**5.991476**
18	***p*-Value**	**0.004603**
19	**Reject the null hypothesis**	

The corrected output is shown below.

	A	B
1	**Kruskal Wallis Rank Test**	
2		
3	**Level of Significance**	**0.01**
4	**Group 1**	
5	**Sum of Ranks**	**94.5**
6	**Sample Size**	**10**
7	**Group 2**	
8	**Sum of Ranks**	**223**
9	**Sample Size**	**10**
10	**Group 3**	
11	**Sum of Ranks**	**147.5**
12	**Sample Size**	**10**
13	Sum of Squared Ranks/Sample Size	8041.55
14	Sum of Sample Sizes	30
15	Number of groups	3
16	*H* Test Statistic	10.76194
17	**Critical Value**	**9.210351**
18	***p*-Value**	**0.004603**
19	**Reject the null hypothesis**	

◀

Technology Lab

▶ Exercise 3 (pg. 549)

Performing a Wilcoxon Rank Sum Test

If the PHStat add-in has not been loaded, you will need to load it before continuing. Follow the instructions in Section GS 8.2.

1. Open worksheet "Tech11_a" in the Chapter 11 folder.

2. At the top of the screen, select **PHStat → Two-Sample Tests → Wilcoxon Rank Sum Test**.

3. Complete the Wilcoxon Rank Sum Test dialog box as shown below. The Population 1 Sample Cell Range is A1 through A13, and the Population 2 Sample Cell Range is B1 through B13. Click **OK**.

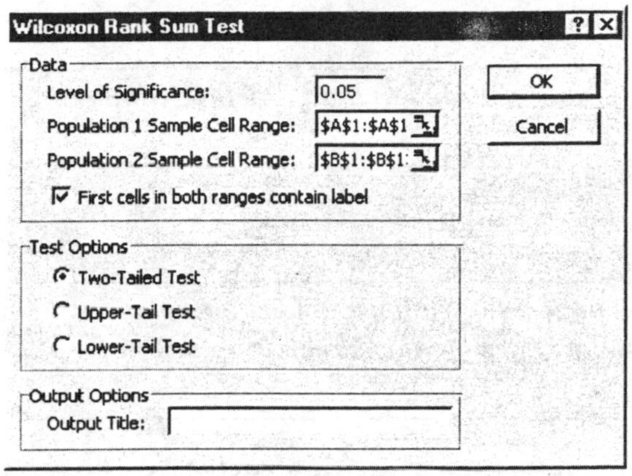

Wilcoxon Rank Sum Test

Data
- Level of Significance: 0.05
- Population 1 Sample Cell Range: $A$1:$A$1
- Population 2 Sample Cell Range: $B$1:$B$1:
- [x] First cells in both ranges contain label

Test Options
- (•) Two-Tailed Test
- ( ) Upper-Tail Test
- ( ) Lower-Tail Test

Output Options
- Output Title:

OK | Cancel

Your output should look similar to the output shown below.

	A	B
1	**Wilcoxon Rank Sum Test**	
2		
3	**Level of Significance**	**0.05**
4	**Population 1 Sample**	
5	**Sample Size**	**12**
6	**Sum of Ranks**	**152**
7	**Population 2 Sample**	
8	**Sample Size**	**12**
9	**Sum of Ranks**	**148**
10		
11	Total Sample Size n	24
12	*T*1 Test Statistic	152
13	*T*1 Mean	150
14	Standard Error of *T*1	17.32051
15	Z Test Statistic	0.11547
16		
17	**Two-Tailed Test**	
18	**Lower Critical Value**	**-1.95996**
19	**Upper Critical Value**	**1.959961**
20	***p* value**	**0.908072**
21	**Do not reject the null hypothesis**	

◀

► Exercise 4 (pg. 549) Performing a Kruskal-Wallis Test

If the PHStat add-in has not been loaded, you will need to load it before continuing. Follow the instructions in Section GS 8.2.

1. Open worksheet "Tech11_a" in the Chapter 11 folder.

If you have just completed Exercise 3 on page 549 and have not yet closed the worksheet, click on the ***Tech11_a*** *tab at the bottom of the screen to return to the worksheet containing the data.*

2. At the top of the screen, select **PHStat → c-Sample Tests → Kruskal-Wallis Rank Test**.

3. Complete the Kruskal-Wallis Rank Test dialog box as shown below. The Sample Data Cell Range is A1 through D13. Click **OK**.

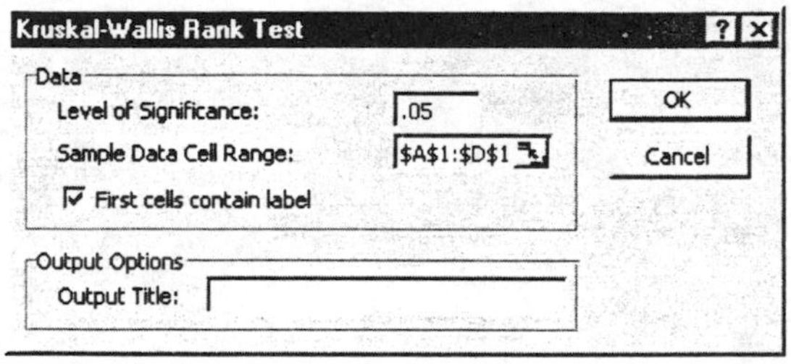

Kruskal-Wallis Rank Test
Data
Level of Significance: .05
Sample Data Cell Range: $A$1:$D$1
First cells contain label
OK
Cancel
Output Options
Output Title:

Your output should appear similar to the output shown below. You may have to scroll down a couple rows to see the entire output.

	A	B
1	**Kruskal Wallis Rank Test**	
2		
3	**Level of Significance**	0.05
4	**Group 1**	
5	**Sum of Ranks**	281
6	**Sample Size**	12
7	**Group 2**	
8	**Sum of Ranks**	256.5
9	**Sample Size**	12
10	**Group 3**	
11	**Sum of Ranks**	227
12	**Sample Size**	12
13	**Group 4**	
14	**Sum of Ranks**	411.5
15	**Sample Size**	12
16	Sum of Squared Ranks/Sample Size	30467.88
17	Sum of Sample Sizes	48
18	Number of groups	4
19	*H* Test Statistic	8.448342
20	**Critical Value**	7.814725
21	***p* Value**	0.0376

◄

► Exercise 5 (pg. 549) — Performing a One-Way ANOVA

1. Open worksheet "Tech11_a" in the Chapter 11 folder.

*If you have just completed Exercise 3 or Exercise 4 on page 549 and have not yet closed the worksheet, click on the **Tech11_a** tab at the bottom of the screen to return to the worksheet containing the data.*

2. At the top of the screen, click **Tools → Data Analysis**. Select **Anova: Single Factor** and click **OK**.

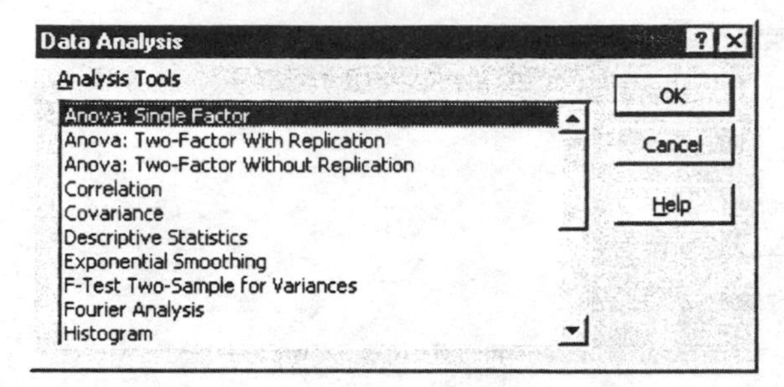

3. Complete the Anova: Single Factor dialog box as shown below. Click **OK**.

Anova: Single Factor
Input
Input Range: $A$1:$D$13
Grouped By: Columns
Rows
Labels in First Row
Alpha: 0.05
Output options
Output Range:
New Worksheet Ply:
New Workbook
OK
Cancel
Help

Adjust the column width so that you can read all the labels. Your output should appear similar to the output shown below.

	A	B	C	D	E	F	G
1	Anova: Single Factor						
2							
3	SUMMARY						
4	*Groups*	*Count*	*Sum*	*Average*	*Variance*		
5	NE	12	1493.9	124.4917	1610.355		
6	MW	12	1393.3	116.1083	485.5845		
7	S	12	1323	110.25	246.67		
8	W	12	1627	135.5833	392.6361		
9							
10							
11	ANOVA						
12	*Source of Variation*	*SS*	*df*	*MS*	*F*	*P-value*	*F crit*
13	Between Groups	4354.512	3	1451.504	2.122667	0.110955	2.816464
14	Within Groups	30087.71	44	683.8115			
15							
16	Total	34442.22	47				

◀

LICENSE AGREEMENT

YOU SHOULD CAREFULLY READ THE FOLLOWING TERMS AND CONDITIONS BEFORE BREAKING THE SEAL ON THE PACKAGE. AMONG OTHER THINGS, THIS AGREEMENT LICENSES THE ENCLOSED SOFTWARE TO YOU AND CONTAINS WARRANTY AND LIABILITY DISCLAIMERS. BY BREAKING THE SEAL ON THE PACKAGE, YOU ARE ACCEPTING AND AGREEING TO THE TERMS AND CONDITIONS OF THIS AGREEMENT. IF YOU DO NOT AGREE TO THE TERMS OF THIS AGREEMENT, DO NOT BREAK THE SEAL. YOU SHOULD PROMPTLY RETURN THE PACKAGE UNOPENED.

LICENSE.

Subject to the provisions contained herein, Prentice-Hall, Inc. ("PH") hereby grants to you a non-exclusive, non-transferable license to use the object code version of the computer software product ("Software") contained in the package on a single computer of the type identified on the package.

SOFTWARE AND DOCUMENTATION.

PH shall furnish the Software to you on media in machine-readable object code form and may also provide the standard documentation ("Documentation") containing instructions for operation and use of the Software.

LICENSE TERM AND CHARGES.

The term of this license commences upon delivery of the Software to you and is perpetual unless earlier terminated upon default or as otherwise set forth herein.

TITLE.

Title, and ownership right, and intellectual property rights in and to the Software and Documentation shall remain in PH and/or in suppliers to PH of programs contained in the Software. The Software is provided for your own internal use under this license. This license does not include the right to sublicense and is personal to you and therefore may not be assigned (by operation of law or otherwise) or transferred without the prior written consent of PH. You acknowledge that the Software in source code form remains a confidential trade secret of PH and/or its suppliers and therefore you agree not to attempt to decipher or decompile, modify, disassemble, reverse engineer or prepare derivative works of the Software or develop source code for the Software or knowingly allow others to do so. Further, you may not copy the Documentation or other written materials accompanying the Software.

UPDATES.

This license does not grant you any right, license, or interest in and to any improvements, modifications, enhancements, or updates to the Software and Documentation. Updates, if available, may be obtained by you at PH's then current standard pricing, terms, and conditions.

LIMITED WARRANTY AND DISCLAIMER.

PH warrants that the media containing the Software, if provided by PH, is free from defects in material and workmanship under normal use for a period of sixty (60) days from the date you purchased a license to it.

THIS IS A LIMITED WARRANTY AND IT IS THE ONLY WARRANTY MADE BY PH. THE SOFTWARE IS PROVIDED 'AS IS' AND PH SPECIFICALLY DISCLAIMS ALL WARRANTIES OF ANY KIND, EITHER EXPRESS OR IMPLIED, INCLUDING, BUT NOT LIMITED TO, THE IMPLIED WARRANTY OF MERCHANTABILITY AND FITNESS FOR A PARTICULAR PURPOSE. FURTHER, COMPANY DOES NOT WARRANT, GUARANTY OR MAKE ANY REPRESENTATIONS REGARDING THE USE, OR THE RESULTS OF THE USE, OF THE SOFTWARE IN TERMS OF CORRECTNESS, ACCURACY, RELIABILITY, CURRENTNESS, OR OTHERWISE AND DOES NOT WARRANT THAT THE OPERATION OF ANY SOFTWARE WILL BE UNINTERRUPTED OR ERROR FREE. COMPANY EXPRESSLY DISCLAIMS ANY WARRANTIES NOT STATED HEREIN. NO ORAL OR WRITTEN INFORMATION OR ADVICE GIVEN BY PH, OR ANY PH DEALER, AGENT, EMPLOYEE OR OTHERS SHALL CREATE, MODIFY OR EXTEND A WARRANTY OR IN ANY WAY INCREASE THE SCOPE OF THE FOREGOING WARRANTY, AND NEITHER SUBLICENSEE OR PURCHASER MAY RELY ON ANY SUCH INFORMATION OR ADVICE. If the media is subjected to accident, abuse, or improper use; or if you violate the terms of this Agreement, then this warranty shall immediately be terminated. This warranty shall not apply if the Software is used on or in conjunction with hardware or programs other than the unmodified version of hardware and programs with which the Software was designed to be used as described in the Documentation.

LIMITATION OF LIABILITY.

Your sole and exclusive remedies for any damage or loss in any way connected with the Software are set forth below. UNDER NO CIRCUMSTANCES AND UNDER NO LEGAL THEORY, TORT, CONTRACT, OR OTHERWISE, SHALL PH BE LIABLE TO YOU OR ANY OTHER PERSON FOR ANY INDIRECT, SPECIAL, INCIDENTAL, OR CONSEQUENTIAL DAMAGES OF ANY CHARACTER INCLUDING, WITHOUT LIMITATION, DAMAGES FOR LOSS OF GOODWILL, LOSS OF PROFIT, WORK STOPPAGE, COMPUTER FAILURE OR MALFUNCTION, OR ANY AND ALL OTHER COMMERCIAL DAMAGES OR LOSSES, OR FOR ANY OTHER DAMAGES EVEN IF PH SHALL HAVE BEEN INFORMED OF THE POSSIBILITY OF SUCH DAMAGES, OR FOR ANY CLAIM BY ANY OTHER PARTY. PH'S THIRD PARTY PROGRAM SUPPLIERS MAKE NO WARRANTY, AND HAVE NO LIABILITY WHATSOEVER, TO YOU. PH's sole and exclusive obligation and liability and your exclusive remedy shall be: upon PH's election, (i) the replacement of your defective media; or (ii) the repair or correction of your defective media if PH is able, so that it will conform to the above warranty; or (iii) if PH is unable to replace or repair, you may terminate this license by returning the Software. Only if you inform PH of your problem during the applicable warranty period will PH be obligated to honor this warranty. You may contact PH to inform PH of the problem as follows:

SOME STATES OR JURISDICTIONS DO NOT ALLOW THE EXCLUSION OF IMPLIED WARRANTIES OR LIMITATION OR EXCLUSION OF CONSEQUENTIAL DAMAGES, SO THE ABOVE LIMITATIONS OR EXCLUSIONS MAY NOT APPLY TO YOU. THIS WARRANTY GIVES YOU SPECIFIC LEGAL RIGHTS AND YOU MAY ALSO HAVE OTHER RIGHTS WHICH VARY BY STATE OR JURISDICTION.

MISCELLANEOUS.

If any provision of this Agreement is held to be ineffective, unenforceable, or illegal under certain circumstances for any reason, such decision shall not affect the validity or enforceability (i) of such provision under other circumstances or (ii) of the remaining provisions hereof under all circumstances and such provision shall be reformed to and only to the extent necessary to make it effective, enforceable, and legal under such circumstances. All headings are solely for convenience and shall not be considered in interpreting this Agreement. This Agreement shall be governed by and construed under New York law as such law applies to agreements between New York residents entered into and to be performed entirely within New York, except as required by U.S. Government rules and regulations to be governed by Federal law.

YOU ACKNOWLEDGE THAT YOU HAVE READ THIS AGREEMENT, UNDERSTAND IT, AND AGREE TO BE BOUND BY ITS TERMS AND CONDITIONS. YOU FURTHER AGREE THAT IT IS THE COMPLETE AND EXCLUSIVE STATEMENT OF THE AGREEMENT BETWEEN US THAT SUPERSEDES ANY PROPOSAL OR PRIOR AGREEMENT, ORAL OR WRITTEN, AND ANY OTHER COMMUNICATIONS BETWEEN US RELATING TO THE SUBJECT MATTER OF THIS AGREEMENT.

U.S. GOVERNMENT RESTRICTED RIGHTS.

Use, duplication or disclosure by the Government is subject to restrictions set forth in subparagraphs (a) through (d) of the Commercial Computer-Restricted Rights clause at FAR 52.227-19 when applicable, or in subparagraph (c) (1) (ii) of the Rights in Technical Data and Computer Software clause at DFARS 252.227-7013, and in similar clauses in the NASA FAR Supplement.